Protein purification techniques

Second Edition

A Practical Approach

Edited by

Simon Roe

AEA Technology plc, Bioprocessing Facility,
Transport Way, Watlington Road,
Oxford OX4 6LY

OXFORD
UNIVERSITY PRESS

*This book has been printed digitally and produced in a standard specification
in order to ensure its continuing availability*

OXFORD
UNIVERSITY PRESS

Great Clarendon Street, Oxford OX2 6DP

Oxford University Press is a department of the University of Oxford.
It furthers the University's objective of excellence in research, scholarship,
and education by publishing worldwide in

Oxford New York

Auckland   Cape Town Dar es Salaam   Hong Kong   Karachi
Kuala Lumpur Madrid Melbourne Mexico City Nairobi
New Delhi Shanghai Taipei Toronto
With offices in
Argentina Austria Brazil Chile Czech Republic France Greece
Guatemala Hungary Italy Japan South Korea Poland Portugal
Singapore Switzerland Thailand Turkey Ukraine Vietnam

Oxford is a registered trade mark of Oxford University Press
in the UK and in certain other countries

Published in the United States
by Oxford University Press Inc., New York

© Oxford University Press, 2001

The moral rights of the author have been asserted

Database right Oxford University Press (maker)

Reprinted 2006

All rights reserved. No part of this publication may be reproduced,
stored in a retrieval system, or transmitted, in any form or by any means,
without the prior permission in writing of Oxford University Press,
or as expressly permitted by law, or under terms agreed with the appropriate
reprographics rights organization. Enquiries concerning reproduction
outside the scope of the above should be sent to the Rights Department,
Oxford University Press, at the address above

You must not circulate this book in any other binding or cover
And you must impose this same condition on any acquirer

ISBN 0-19-963673-7

Protein purification techniques

The Practical Approach Series

Related **Practical Approach** Series Titles

* indicates a forthcoming title

Please see the **Practical Approach** series website at
http://www.oup.co.uk/pas
for full contents lists of all Practical Approach titles.

Preface

Biotechnology is now playing an increasingly important role in our lives. The last 20 years has seen an explosion in the application of genetics, cells, and their components to improve the quality of life, from new therapeutic treatments and rapid diagnostic tests to improved disease resistance in crops. At the time of writing, some 1300 biotechnology companies have emerged in the USA employing over 150 000 people, with new therapeutic products providing multi-million dollar sales. This same trend is now being repeated in Europe and the next five years should see the successful commercialization of new biotechnology companies and the transition of biotechnology start-ups into established players.

Such commercial progress is, of course, the result of our improved understanding of the life sciences and the causes of human disease. While new developments in cell biology, immunology, and genetics have been widely publicized, the significant advances in separations technology have played a crucial role in improving our knowledge of the life sciences. Since the early days of biochemistry, when the fundamentals of cell metabolism and enzyme function were being established, protein separation techniques have been an essential part of the biotechnology revolution. Any biochemistry laboratory, whether in schools, universities, high-tech biotechnology companies, or pharmaceuticals giants, will invariably contain the essential tools of the trade—centrifuges, spectrophotometers, buffers, and chromatography columns. In fact proteins are perhaps the most widely purified type of biological molecule since they form an integral part of cellular metabolism, structure, and function. Protein purification has continued to evolve during the last ten years, with improvements in equipment control, automation, separations materials, and the introduction of new protein separation techniques such as affinity membranes and expanded beds. Microseparation techniques have emerged for protein sequencing and the highly sophisticated (and often expensive) equipment now available allows chromatography process development in hours where days or even weeks might have been required in the past.

These advances in protein separation are likely to continue, driven by the need for faster process development and improved resolution. Such developments have clearly reduced the workload but may have removed the purification scientist from an understanding of the fundamentals of the science of purification. Although sophisticated equipment may now ease the workload, there is still a need to consider how purification strategies are designed and unit operations joined together to produce a processing train which can, if required, be scaled-up efficiently. In addition, an understanding of protein purification facilitates problem solving and enables the design of a robust process for production purposes.

As a result, it is appropriate that two new editions of 'Practical approaches in protein purification' should appear, covering protein purification techniques and protein purification applications. The focus of these publications continues to be on the provision of detailed practical guidelines for purifying proteins, particularly at the laboratory scale and, where possible, information is summarized in tables and recipes. The books are primarily aimed at the laboratory worker and assumes an understanding of the basics of biochemistry. Such are the advances in protein purification and the widespread use of the techniques that the two volumes cannot be comprehensive. Nevertheless each of the key techniques used at a laboratory scale is covered, starting from an overview of purification strategy and analytical techniques, and followed by initial extraction and clarification techniques. Since chromatography forms the backbone of modern purifications strategies, several chapters are devoted to this technique. As with the previous editions, several chapters also cover specific considerations involved in the purification from certain protein sources. While the emphasis of the books are on laboratory scale operations, where relevant, information is included within the appropriate chapter on scale-up considerations, and one is devoted to this area in more detail. Given the wide variety of knowledge involved, each chapter has been written by an author who is a specialist on the given subject.

Although successful protein purification requires an understanding of both biochemistry and the science of separation, I have always thought that the process of designing a purification strategy has an artistic element involved and is a challenge to be enjoyed rather than laboured over. Starting from a crude mixture of proteins, carbohydrates, lipids, and cell debris, a well thought out sequence of steps leading ultimately to a highly pure single protein is a work of art which must be admired. I hope that the new editions kindle the intense interest in protein purification which I have enjoyed over the years. Who knows, even those who have long hung up their laboratory coats may head down to the labs in the late hours to potter around with the odd ion exchange column. My apologies to their partners in advance!

Didcot   S. D. R.
November 2000

Contents

CONTENTS

Protocol list

Abbreviations

AIEC	anionic exchange
ATP	adenosine triphosphate
BCA	bicinchonic acid
Bis-Tris	bis-(2-hydroxyethyl) imino-Tris-(hydroxyethyl) methane
BSA	bovine serum albumin
CDI	N,N' carbonyldiimidazole
CIEC	cationic exchange
CIP	cleaning in place
CNBr	cyanogen bromide
DMEM	Dulbecco's minimal Eagle's medium
DNA	deoxyribonucleic acid
DTT	dithiothreitol
EDTA	ethylenediaminetetraacetic acid
EGTA	ethylene glycol bis (β-aminoethyl ether) N,N,N',N',-tetraacetic acid
ELISA	enzyme-linked immunosorbent assay
FDA	Food and Drug Administration
FITC	fluorescein isothiocyanate
FMN	flavin monophosphate
FMP	2-fluoro-1-methyl pyridinium toluene-4-sulfonate
HBsAg	hepatitis B virus surface antigen
Hepes	N-2-hydroxyethylpiperazine-N-2-ethane sulfonic acid
HIC	hydrophobic interaction chromatography
HPLC	high performance liquid chromatography
IDA	iminodiacetic acid
IEC	ion exchange chromatography
IEF	isoelectric focusing
IgG	immunoglobulin G
IMAC	immobilized metal affinity chromatography
LPS	lipopolysaccharides
mAbs	monoclonal antibodies
Mes	N-morpholinoethane sulfonic acid
Mops	N-morpholinopropane sulfonic acid
NMWC	nominal molecular weight cut-off

NTA	nitrilotriacetic acid
OD	optical density
PBA	phenyl boronate
PBS	phosphate-buffered saline
PEG	polyethylene glycol
PMSF	phenylmethylsulfonyl fluoride
PVPP	polyvinylpolypyrrolidone
RDVF	rotary drum vacuum filters
RFC	radial flow chromatography
RPC	reversed-phase chromatography
SDS	sodium dodecyl sulfate
SDS–PAGE	sodium dodecyl sulfate–polyacrylamide gel electrophoresis
SEC	size exclusion chromatography
TBS	Tris-buffered saline
TCA	trichloroacetic acid
TEMED	N,N,N',N',-tetramethyethylenediamine
Tosyl	p-toluene sulfonyl
Tresyl	trifluoroethyl sulfonyl
Tricine	N-Tris (hydroxymethyl) methyl-glycine
Tris	Tris (hydroxymethyl) aminomethane
UV	ultraviolet

Chapter 1
Purification strategy

Simon D. Roe

AEA Technology plc, Bioprocessing Facility, Transport Way, Watlington Road, Oxford OX4 6LY

Practical approaches to protein purification contains detailed practical information on separations techniques and the chapters of *Protein purification techniques* cover the unit operations and analytical techniques involved in some detail while *Protein purification applications* provides details of how to approach purification from a selected number of typical sources. However a key element of every purification, whether in University research or as part of scaling-up an industrial process, is planning. Time spent at the outset in establishing the key goals of the process prior to going into the laboratory will invariably save much time and effort. Key points to consider from the outset are:

(a) Why are you doing the work—what is the reason for the purification?

(b) What are the key considerations in selecting how to purify?

(c) What implications does this have on how you will approach the purification?

1 Key considerations

At the start of a purification, the target protein may be a minor component among millions of other proteins and other contaminants. This presents boundless opportunities for miscalculations, blind-alleys, and wasted effort. Years of practical work by separation scientists have derived certain rules which will help to minimize such problems and ensure protein purifications are successful. Here are ten purification rules to consider:

1. Keep the purification simple—minimize the number of steps and avoid difficult manipulations which will not reproduce.

2. Keep it cheap—avoid expensive techniques where a cheaper one will do.

3. Adopt a step approach—and optimize each step as you go.

4. Speed is important—avoid delays and slow equipment.

5. Use reliable techniques and apparatus.

6. Spend money on simple bits and pieces—e.g. test-tubes, pipettes.

7. Write out your methods before you start and record what you have done accurately.

8. Ensure your assays are developed to monitor the purification.

9. Keep notes on yields and activity throughout.

10. Bear in mind your objectives—be it high yield, high purity, final scale of operation, reproducibility, economical use of reagents/apparatus, convenience, throughput.

2 Aims of purification—why?

Protein purification uses time, money, effort, and valuable equipment. Therefore, it is always advisable to pause for some moments to consider the reasons for purification in the first place. For many this is blindingly obvious since you are immersed in a project in which the goal is well-defined. The golden rule to which every purification must adhere to is:

'*never purify more than is required by the end use*'.

Typical reasons for purifying a protein may be:

(a) To identify the function of the protein (e.g. an enzyme).

(b) To identify the structure of the protein.

(c) To use the enzyme to generate a desired product as part of a research project.

(d) To produce a commercial product such as a diagnostic or therapeutic.

At a research level, in the laboratory, extensive purification will always yield additional information about a particular protein. However, in industrial projects where scale-up to provide a manufacturing route is required, purification for purpose becomes paramount.

At one extreme we may require a bulk enzyme such as amylase which is used in brewing and baking. This is a low cost product where removal only of bulk impurities such as cells and cell debris, concentration, and some formulation are required. The purification is consequently short, and cheap. Only those impurities which interfere with the action or stability of the enzyme, or which are restricted in its final application, are removed. Such an approach is also required at the laboratory scale where enzyme function is being elucidated and interferences to enzyme activity must be removed.

At the other extreme may be a therapeutic product where achievement of 99.9% purity through removal of most impurities is essential to minimize side-effects following injection, and regulatory bodies such as the FDA must be satisfied that the final product is safe for its designated use. Here a lengthy purification is probably required, with several unit operations such as chromatography necessary to reduce to acceptable concentrations any undesirable impurities such as endotoxin, DNA, and virus particles. Such purity is not only required in therapeutics manufacture—research for identification of protein structure requires crystallization which in turn demands a high purity product, albeit at only small scales of operation.

Having established the reasons for carrying out a purification, it is advisable to spend some time in planning your approach to the task and gathering what information is already available concerning the proposed project.

3 The target protein and contaminants

Knowledge of the structure, function, and properties of the protein which is to be purified is very important and plays a central role in designing the purification strategy (1). Information on molecular weight, isoelectric point, hydrophobicity, presence of carbohydrate, affinity for substrates, and sensitivity to metal ions can help in establishing the key steps in a purification. What is known of the stability of the protein at varying pH and temperature, and following exposure to organic solvents, heavy metal ions, shear, and proteases? A key source of such information is the literature. Of particular interest here is any unusual properties which can be used to advantage in the purification. As an example if the protein is unusually thermostable, then raising the temperature can be used to denature contaminating proteins to leave the target protein relatively pure. Such drastic purification steps can be used early on in a process to great benefit, and although not 'high tech' compared to modern purification techniques, they should always be borne in mind as a potentially great timesaver.

If no information can be found in the literature, then some simple laboratory tests can help in establishing this data. As an example, simple packed column tests will help to establish protein charge or size by using ion exchange or size exclusion, respectively. Simple parallel columns can be used to speed up this evaluation stage.

At this point, any knowledge of contaminants present should be accumulated. Purification strategies can be designed to provide effective removal of contaminants from the target protein as well as vice versa. The types of contaminants will depend very much on the source of the target protein (e.g. fermentation, animal tissue, plant material) and its location within that material (e.g. extracellular versus intracellular). Typical contaminants will include nucleic acids, lipids, other protein, cells, cell debris, and pigments such as polyphenols (*Table 1*). As an example, nucleic acids may cause a high viscosity following bacterial cell lysis, so reducing the efficiency of purification steps and sample handling. Addition of nuclease enzymes to the feedstock prior to cell lysis can be used to hydrolyse the nucleic acids into short chains and so reduce viscosity, and improve sample handling. In adopting this approach it is essential to remember that contaminants should only be removed if they interfere with the end use of the protein, or prevent efficient handling during the purification (as with nucleic acids in the example described here). In general it is best to purify so as to optimize yield and specific activity rather than for eliminating each type of contaminant.

Table 1 Typical classes of contaminants found in protein purification

Particulates	Include cells and cell debris. Usually removed initially using centrifugation or filtration.
Proteins	Include general host cell protein and proteins of similar properties to the target protein. Gross protein contaminants removed using precipitation while adsorption and chromatography techniques are widely applicable.
Modified target protein	Target protein modified through altered amino acid sequence, glycosylation, denaturation, etc. Removed using chromatography.
Lipids, lipoproteins	May be derived from host cells (e.g. membranes) or added to a fermentation (e.g. antifoams).
Small molecules	Include salts, sugars and reagents added to a purification. Typically removed using gel permeation chromatography or diafiltration.
Polyphenols	Coloured compounds often derived from plant sources, and often included in crude fermentation ingredients such as corn steep liquor. Removed by precipitation or chromatography.
Nucleic acids	Released during cell lysis and may increase sample viscosity. Removed using ion exchange, precipitation techniques such as protamine sulfate, or through hydrolysis with nucleases.
Pyrogens	Usually lipopolysaccharides derived from Gram negative bacterial cell walls.
Aggregates	Include inclusion bodies which are solubilized using chaotropic agents such as 6 M urea. Aggregate contaminants may be removed using gel permeation chromatography.

4 Protein structure

All proteins are composed of amino acids which are linked together into a peptide chain via peptide bonds. There are 20 amino acids which form the basic building blocks of proteins, each with a characteristic side group which determines the properties of the amino acid. Proteins are composed of one or more peptide chain in a three-dimensional (tertiary) structure which may also include linkages with non-protein molecules such as lipids and carbohydrates. The combination of number and type of amino acids and the non-protein components of the tertiary structure determine the properties of the protein. The three-dimensional configuration is often stabilized by both covalent and non-covalent interactions and disruption of this conformation (for example through heating or a low pH) causes denaturation and loss of any activity which may be associated with the protein. The properties of a protein (*Table 2*) are determined by the chemical groups which are exposed on the surface and the overall molecular weight of the molecule, which determines the Stokes radius. Important chemical properties include charge, hydrophobicity, solubility, and biological activity. The overall properties of a protein in relation to contaminants which are present will determine the most appropriate techniques to be used for a purification.

Table 2 Important properties of proteins and their relevance to purification

Charge	A measure of the surface ionic properties of a protein, giving the molecule a net negative or positive charge. Charge varies with pH and the isoelectric point is the pH at which the net charge is zero.
Biological activity	The presence of sites on a protein which can interact with other biological molecules with a high degree of affinity. Such interactions can be used to purify the target protein.
Hydrophobicity	A measure of the water-hating character of a protein. Highly hydrophobic molecules such as membrane proteins may be insoluble without addition of solvents or detergents. Highly hydrophobic proteins also precipitate at low concentrations of ammonium sulfate.
Size	Molecular weight is usually measured by gel electrophoresis (SDS–PAGE). The Stokes' radius, which gives an indication of shape.
Solubility/stability	The ability of a protein molecule to remain in solution is influenced by pH, salt concentration, presence of solvents and detergents, hydrophobicity, and biological affinity.

5 Source of the product

Proteins are purified from a wide variety of sources, including blood, animal tissue (e.g. heart, muscle, and liver), plant tissue (spinach, horseradish), and fermentation. Volume 2 (*Protein purification applications*) considers the influence of source material on purification strategy in some detail. In natural sources such as organs and bodily fluids a protein may be present in very low concentrations such that purification is a challenging task. Consequently, the genetic information coding for a protein is often transferred into recombinant cells which can be cultured to produce the target protein in significantly increased yields. Bacteria, yeast, and insect cells are all used for this purpose, with transgenic animals also being used as vehicles for increased production of proteins.

A common source is fermentation and cell culture with yeast, fungi, bacteria, animal, and plant cells all commonly used. Although it is likely that you will not have any choice over the starting material for your purification, it must be appreciated that the choice of protein source can have a significant impact on the ease of purification. It is wise to check the literature for alternative sources of your protein and, if necessary, evaluate alternative sources for differences in protein yields, specific activity, and stability. One should not assume that if a method described in the literature uses one particular source of a protein, that this source is the optimum starting material. Widen your literature search and see if you can uncover any other options and pay particular attention to noting down initial product titres, final yields, and purities which are obtained from different sources. It is also important to realize that biological materials are often not reproducible and raw materials such as plant and animal tissues can produce wide variations in the yield of a particular protein. Any starting material such as plant and animal tissue must readily be available in a fresh stock—old material may have lost the required protein through degradation processes, in which case impure purifications will inevitably result. Well-characterized and optimized fermentations provide the opportunity for reproducible starting material and the potential for significantly improved protein yields through genetic manipulation, strain selection, and culture optimization. Practical aspects of selecting and handling starting material are described in more detail in Chapter 2.

6 Key steps in purification

While no two purifications are the same, and the separation scientist is faced with a bewildering variety of techniques available to achieve their goal, the majority of purifications aim to achieve the same result. These are:

- release of target protein from the starting material
- removal of solids to leave the protein in a supernatant
- removal of water to concentrate the protein
- removal of contaminants to achieve the desired purity
- stabilization of the target protein

These five goals form the basis of the stages used in protein purification as follows:

(a) **Stage 1: Initial fractionation**. Here the goal is to prepare the protein as a clarified solution. This typically involves techniques such as centrifugation, microfiltration, and cell lysis. Concentration and coarse fractionation may also be achieved using techniques such as salt precipitation and ultrafiltration to reduce the volume to a manageable amount for subsequent handling.

(b) **Stage 2: Purification**. Initial fractionation is followed by a sequence of operations designed to remove contaminants and concentrate the product further. Chromatography is usually the work-horse of this stage and protein purifications may use three to five chromatography steps in sequence to remove different types of contaminants at each step.

(c) **Stage 3: Polishing**. Although stability of the product throughout the process is important, the protein will probably be stored for a period after the purification and so generation of optimum conditions for protein stability is usually a key goal of the final stages of the purification process. The removal of aggregated or degraded protein is also important at this stage, using size exclusion chromatography as a key step.

7 Process integration

A well-designed protein purification must not only have optimized steps to achieve the goals described above, but these steps must be linked together into a sequence which allows the product of one step to be fed into the next step with minimal adjustment. This is particularly important where a process is to be scaled-up. Unit operations such as centrifugation, salt precipitation, and chromatography may be optimized in isolation but they should always be selected with a view to the purification as a whole. It is important that one step produces material which is in a suitable form to be applied to the next step in the process —pH, salt concentration, and presence of particulates are important considerations.

As an example, chromatography using packed beds requires the protein sample to be free of particulates so as to avoid bed clogging and build up of back-pressure. This requires that the sample is clarified to remove all particulates using centrifugation or filtration techniques. When using a sequence of chromatography steps considerations of sample composition become paramount. A protein solution with the wrong pH or salt conductivity will result in suboptimal chromatography and likely loss of product.

8 Scale-up

In many cases the development of a purification at a laboratory scale will form the first step in a scale-up which may lead to a fully commercial process involving the handling of thousands of litres of product. Decisions made at the small scale will, therefore, form the basis of the full-scale operation and although changes are possible at a later stage it is wise to develop a laboratory purification with a view as to how easy it will operate in a process plant. While process scale-up is discussed in detail elsewhere, certain points are important at the planning stage:

(a) Keep volumes as small as possible to assist in scaling-up. The volume of your product solution determines the size of equipment required for its processing and the total time required for the process. This has implications on cost of equipment, labour, and total time required. Always reduce volume as much as possible as early as possible.

(b) Scaling-up often increases time for processing. If you are handling a labile product this may result in unacceptably low yields of protein and a non-viable process. Always consider options for quicker ways of carrying out a process.

(c) Minimize the handling operations between steps. While sample adjustments using operations such as dilution, filtering, and centrifugation are easy at a small scale, they all add to the time and cost required at a large scale. Try to order the unit operations used in a process such that there is minimal sample adjustment and handling between steps.

(d) Certain unit operations (e.g. centrifugation and homogenization) may not be reproducible as they are scaled-up. In centrifugation scale-up may use a lower g force so that precipitates and cell pastes are less solid and contain more contaminants. Be prepared to adjust such unit operations (e.g. addition of an extra filtration step to remove fine particulates after scale-up of centrifugation).

9 Monitoring the purification—assays

Assays form a very important aspect of any successful purification and are described in detail in Chapter 3. Within the context of the overall purification strategy, you will only know how good your purification is if you have well-developed assays with which to monitor progress. Therefore, it is critical to develop your assays early on so as to monitor procedures and to bear in mind the required sensitivity, accuracy, and speed, within the budget limitations of the project. The two essential assay types required in purification monitoring are for total protein present and for the target protein.

Total protein is most commonly assayed using UV absorption at 280 nm, a result of the absorption of ultraviolet light photons by electrons in aromatic amino acids such as tyrosine, phenylalanine, and tryptophan. While the peptide bonds of proteins cause absorption at 210 nm this wavelength is rarely used for total protein measurement due to the interference from other chemicals which absorb UV light at this wavelength, particularly those containing double bonds between carbons or carbon and oxygen (2). When using 280 nm to measure total protein concentration it should be appreciated that plastic containers which are used to store buffers may leach UV absorbing plasticisers which will interfere with the assay. Furthermore, conditions which alter the tertiary structure of proteins, such as pH and salt concentration, may alter the UV absorption of the protein. Total protein assays using absorption at 280 nm are commonly used to monitor purifications since they are simple and quick to carry out and require a routine spectrophotometer and cuvettes for measurements. Other techniques for the measurement of total protein are discussed in Chapter 3.

Assays for activity are common in enzyme purifications and involve the addition of the enzyme preparation to a substrate in a buffered solution at standardized temperature and pH. The enzyme concentration may be measured by the rate of the catalysed reaction or by the total amount of product synthesized as a result of the enzyme action on the substrate. In the latter case the enzyme reaction is normally allowed to proceed for a predetermined time before the reaction is 'killed' by the addition of a terminating reagent such as acid or alkali to alter the pH and stop the enzyme reaction (3). Such activity assays are important throughout a purification and must be developed and reliable even at the stage of screening for alternative protein sources. At this stage, where the protein may be located within cells, the complete release of all activity from the source material must be achieved before realistic estimates of yields can be made.

10 Time and temperature

Some proteins, particularly those derived from within cells, are labile and literally fall to bits during the purification process. Such loss of product may occur within a few hours if a protein is incorrectly stored. The loss of activity of enzymes is both time- and temperature-dependent—a lengthy purification of a labile protein carried out at room temperature is likely to run into trouble. Consequently it is important to work quickly and if the protein is highly labile, reduce the temperature by working in a cold room.

At the outset, it is therefore important to accumulate what information is available on the lability of your target protein. A literature search will help with this but you can deduce much from the natural conditions from which the protein is derived. As an example, extracellular enzymes from bacterial sources, such as amylases, work in a natural environment which is harsh. Such proteins are likely to be robust and non-problematical in terms of lability. Intracellular enzymes work in an environment which is low in oxygen tension and in which reducing compounds stabilize the enzyme. When released into an oxygen-rich and dilute supernatant they are likely to be labile and require careful handling. What is the temperature at which the protein is normally found? Enzymes derived from bacteria found at elevated temperatures such as hot springs are likely to be perfectly stable at room temperature.

If you are handling a labile protein such as an intracellular enzyme, it is common practice to work at low temperatures. Reducing the temperature from room temperature to 4°C will reduce the rate of degradation processes by a factor of between three and five. However, besides being rather uncomfortable, reducing the operating temperature for a purification also slows down any separation processes. This is because separation processes involving proteins are generally diffusion controlled and so heavily dependent on both temperature and viscosity. As an example, chromatography uses buffers which will have a higher viscosity in the cold due to their water content, so limiting the flow rate which can be applied to the column.

The general rule is that if it is possible to work at room temperature without losing unacceptable levels of activity, then it is best to do so. The working environment is more comfortable and you will be able to achieve far more in a working day and possibly carry out several purification steps to generate the target protein in a more acceptable form for overnight storage than would have been achievable when working at a slower pace in the cold. Conversely, if you are purifying a labile protein where yields are critical and working in the cold achieves a higher retention of activity despite the longer time required for separations, then work should be carried out in the cold room. Generally, inactivation processes are slowed down at low temperature such that it is an advantage to use a low temperature if this is a problem.

Reduced temperatures, including freezing, are also used for the storage of proteins in between purification steps (see Chapter 2). It is unlikely that a whole purification will be achieved in a single day and so appropriate stages for protein storage should be built into any purification strategy. Should reduced temperatures be required for either a purification or for storage then it must be appreciated that the pH of buffers will alter with pH and, where possible, checks made on buffer pH at 4°C.

This chapter has discussed some of the key factors which should be considered prior to carrying out a purification, including the reasons for a purification, typical pitfalls which should be avoided, and points to consider relating to the nature of the product and the starting material. The following chapter provides initial practical guidance on setting up for purification work.

References

1. Janson, J.-C. (1989). In *Protein purification: principles, high resolution methods, and applications* (ed. J.-C. Janson and L. Ryden), p. VI. VCH Publishers Inc.
2. Stoscheck, C. M. (1990). In *Methods in enzymology* (ed. M. P. Deutscher), Vol. 182, pp. 50–68. Academic Press.
3. Rossomando, E. F. (1990). In *Methods in enzymology* (ed. M. P. Deutscher), Vol. 182, pp. 38–49. Academic Press.

Chapter 2
Getting started

Simon D. Roe

AEA Technology plc, Bioprocessing Facility, Transport Way, Watlington Road, Oxford OX4 6LY

This chapter provides a more detailed overview of the equipment requirements for protein purification and the practical aspects of developing a purification strategy, including the ordering of unit operations, buffer preparation, and approaches to minimizing yield losses.

1 Overview of lab equipment

A well-equipped laboratory is an essential prerequisite to successful protein purification and time spent in deciding needs and purchasing essential items will help to avoid panics mid-way through a protocol when a key piece of equipment is lacking. This said, one should avoid expensive purchases (unless budget is no object) as the majority of purifications can be achieved with fairly routine equipment. The few essential pieces of expensive equipment are a spectrophotometer, a centrifuge, and a chromatography set-up. In general, money is best spent in purchasing plenty of the cheaper items such as tubes, beakers, measuring cylinders, salts, and buffers.

Chromatography equipment is an essential item for any purification laboratory. There are plenty of expensive chromatography set-ups available which can provide a remarkable saving in time for process development purposes. If you are not familiar with such equipment and are new to the world of protein purification, I suggest you make do with a simple chromatography set-up to start with, until you become familiar with the technique and learn more about your exact requirements and how alternative tailor-made process development kits differ.

A protein purification laboratory should be equipped with supplies of tap-water, de-mineralized water, and distilled water. Electricity and sinks are taken for granted. Required equipment can be roughly grouped into three categories:

- those for ancillary purposes
- those for detection
- those for separation

Table 1 lists the essential items for the protein purification laboratory.

Table 1 Recommended equipment and materials for the protein purification laboratory

Spectrophotometer with chart recorder
Gel electrophoresis, isoelectric focusing equipment
Refrigerated centrifuge
Bench-top centrifuge
Homogenizers
Chromatography set-up (pumps, gradient mixer, columns, UV detector, chart recorder)
Balances, pH meters, magnetic stirrers
Ice machine
Graduated cylinders
Pipettes (adjustable with disposable tips 5 µl to 5 ml
Beakers
Chromatographic media
Ammonium sulfate
Buffers
Dialysis tubing
Salt
Stabilizing reagents

1.1 Ancillary equipment

Time and money spent wisely on buying adequate supplies of support materials will pay dividends. There is nothing more irritating than having to rush around in search of a clock or some salt half-way through a delicate purification with your enzyme degrading in front of your eyes! Key requirements are tubes, beakers, pipettes, stirrers, and timers. In addition essential chemicals include salts and buffers. Adjustable pipettes (e.g. Gilson) are essential and those suitable for sample volumes from 10 µl to 5000 µl are recommended. A freezer and fridge are also essential, while an ice machine is desirable. Freezers should chill to $-20\,°C$, although a deep freeze providing around $-70\,°C$ may also be needed. Magnetic stirrers with a built-in hot-plate are preferred.

1.2 Detection and analysis equipment

While analytical techniques are discussed in detail in Chapter 3, the broad requirements will be covered briefly here. The spectrophotometer is perhaps the single most important piece of analytical equipment in the purification laboratory. This should allow measurements of absorption in the ultraviolet (UV) and visible wavelength range, preferably between around 190 and 800 nm. Common uses are in total protein determination (e.g. absorption at 280 nm) and in specific assays such as enzyme kinetics measurements. Many spectrophotometers can be linked up to a chart recorder for recording enzyme kinetics assays. An ample supply of disposable cuvettes for visible wavelength measurements, and a small supply of quartz cuvettes for UV wavelength measurements are needed.

The second essential item is a gel electrophoresis set-up. Electrophoresis is the principal method for generating information on composition of a protein sample, including the approximate molecular weight and isoelectric point of the target protein and main contaminants. Sodium dodecyl sulfate–polyacrylamide gel electrophoresis (SDS–PAGE) and isoelectric focusing (IEF) are the two commonly used techniques for determining approximate molecular weight and isoelectric point respectively. Complete gel electrophoresis systems are typically purchased from laboratory suppliers and normally use vertical gel slabs between two glass plates, although ready-made horizontal gels are also available.

1.3 Separations equipment

The exact equipment requirements for separation will depend on the particular application. However the most likely requirements are a centrifuge, equipment for cell lysis, and a chromatography set-up.

Centrifuges are commonly used for separating cells and cell debris from supernatant and for settling precipitates. Common models are refrigerated and spin at up to 20 000 r.p.m. A selection of rotors is essential to accommodate a variety of sample sizes and centrifuge tubes for sample volumes of between 5–500 ml are recommended. Should fractionation of subcellular components such as organelles be anticipated then an ultracentrifuge capable of spinning at up to 80 000 r.p.m. will be required. Bench-top centrifuges are very useful and have a fixed or swing-arm rotor which typically takes 10 ml centrifuge tubes. Microcentrifuges are also useful for sample volumes of 1 ml and less.

The type of cell lysis equipment required will depend on the source material being used for protein purification. Blenders are commonly used for animal and plant tissue while bead mills are more suitable for microbial cell lysis. Ultrasonication equipment is also useful at a laboratory scale.

Chromatography is so extensively used in modern protein purification that money spent in purchasing reliable equipment will pay dividends. The essential requirements are a variety of column dimensions, a pump, fraction collector, UV monitor, chart recorder, tubing, valves, and a selection of gels and resins. Chromatography columns come in varying degrees of complexity and ideally several columns of varying bed volume from 1 ml up to around 200 ml should suit most applications. It is important to be able to vary the column volume using an adjustable plunger.

A peristaltic pump will suffice for most applications, although dual piston pumps are available with many automated chromatography set-ups, providing a more pulse-free delivery of mobile phase. A reliable fraction collector is essential and extra money spent on this piece of equipment is definitely worthwhile since it is the one piece of equipment which seems to go wrong during overnight purifications—a pool of sample on the bench top being the evidence which is found the following morning. Should extensive protein purification be anticipated, it is worthwhile considering a tailor-made automated chromatography set-up (e.g. Bio-Rad, Pharmacia). When using elution gradients, two peristaltic

pumps and a simple stirrer for gradient formation will be required and automated chromatography apparatus is normally fitted with a mixer for gradient
formation. Ion exchange is the single most common chromatography medium
used in protein purification, although additional media such as size exclusion
and hydrophobic interaction are also likely to be needed.

2 Control of pH—buffers—principles, selection, and use

Proteins are pH-sensitive molecules and their stability, and possibly activity,
may be dependent on a narrow pH window. Many enzyme-catalysed reactions
use up or release protons, so causing a shift in pH. In addition biochemical
reactions are generally sensitive to quite small changes in pH. Similarly, the
behaviour of proteins during purification is often highly pH-dependent and
requires that the pH of a protein sample is known at all times, and, if necessary,
adjusted. Buffers are essential to control pH and the protein purification scientist
should be familiar with the common types of buffers used in purification, their
characteristics, and the range of pH in which they can be used. The important
criteria for choosing a buffer are shown in *Table 2*.

Table 2 Key criteria in buffer selection

pK_a
Variation in the pK_a with temperature
Ionic strength required
Form of the buffer is anionic or cationic
Multiple charges
Solubility
Interaction with other components
Cost
UV absorption

Buffers maintain the pH by absorbing and releasing protons and are essential
in protein purification to ensure that proteins are not denatured due to a shift in
pH, and that purifications can be carried out reproducibly under optimum conditions of pH and ionic strength. Buffers consist of an acid and its conjugate base
(e.g. acetic acid and acetate) and are formed by mixing the appropriate amount
of each form of the buffer to achieve the required pH. It is recommended to
keep a concentrated stock of buffer (e.g. a 100-fold concentrate) which is diluted
when required and then the pH fine-adjusted prior to use. The concentrate
should contain a bactericidal agent such as sodium azide to prevent microbial
growth and which will be diluted out when the buffer is prepared for use.

A list of commonly used buffers and their pH windows is shown in *Table 3*.
Buffers stabilize pH most effectively at their pK value, but can generally be used
within 1 pH unit on either side. In selecting the most suitable buffer it is im-

Table 3 Useful buffers and pK_a values at 25°C

Buffer name	pK_a value
Lactic acid	3.86
Acetic acid	4.76
Pyridine	5.23
Succinic acid	5.64
Histidine	6.0
N-morpholinoethane sulfonic acid (Mes)	6.15
Bis-(2-hydroxyethyl)imino-Tris-(hydroxyethyl) methane (Bis-tris)	6.5
Imidazole	6.95
Phosphate	7.2
N-morpholinopropane sulfonic acid (Mops)	7.2
N-2-hydroxyethylpiperazine-N-2-ethane sulfonic acid (Hepes)	7.48
Triethanolamine	7.75
Tris(hydroxymethyl) aminomethane (Tris)	8.06
N-Tris(hydroxymethyl)methyl glycine (tricine)	8.15
Diethanolamine	8.9
Ammonia	9.25
Boric acid	9.23
Ethanolamine	9.5
Glycine	9.8
Carbonate	10.3
Piperidine	11.12

portant that it does not interfere with enzyme function—phosphate buffers, for example, inhibit many enzymes, such as urease, kinases, and dehydrogenases. In addition, Tris does not buffer well below pH 7.5 and its buffering capacity is very temperature-dependent.

Buffer pH is temperature-sensitive and the pH should always be checked at the temperature at which the buffer will be used. In addition, after addition of a buffer to a protein solution, the pH should always be re-checked since proteins are also buffers and some alteration in pH may occur. Buffers are typically used at a 10–50 mM concentration, with Tris, phosphate, and acetate being most commonly used.

A buffer is only as good as the pH meter which is used for its preparation. Consequently, all pH meters should be regularly calibrated using commercially available buffer standards, commonly of pH 5, 7, and 10. It is good practice to ensure that all staff working in a purification laboratory follow the correct, and the same, procedure for pH meter calibration and note down in an equipment log when a meter has been calibrated. It is useful to nominate a member of staff to carry out such calibrations on all pH meters at a regular basis.

Since Tris and phosphate are two of the most commonly used buffers in protein purification, recipes for the preparation of these buffers at varying pH are given in *Protocol 1* (1).

Protocol 1

Preparation of phosphate and Tris buffers—standard recipes

Reagents

- Sodium dihydrogen phosphate
- Disodium hydrogen phosphate (7H$_2$O)
- Tris(hydroxymethyl)aminomethane
- HCl

A Phosphate buffer

1 Preparation of 0.2 M stock solutions:

(a) Stock A: add 27.8 g of sodium dihydrogen phosphate in 1 litre of water.

(b) Stock B: add 53.65 g of disodium hydrogen phosphate (7H$_2$O) in 1 litre of water.

2 Mix x ml of stock A with y ml of stock B, then dilute to a total of 200 ml for the following pH of buffer:

x	y	pH	x	y	pH
87.7	12.3	6.0	39.0	61.0	7.0
85.0	15.0	6.1	33.0	67.0	7.1
81.5	18.5	6.2	28.0	72.0	7.2
77.5	22.5	6.3	23.0	77.0	7.3
73.5	26.5	6.4	19.0	81.0	7.4
68.5	31.5	6.5	16.0	84.0	7.5
62.5	37.5	6.6	13.0	87.0	7.6
56.5	43.5	6.7	10.5	90.5	7.7
56.5	43.5	6.7	8.5	91.5	7.8
51.0	49.0	6.8	7.0	93.0	7.9
45.0	55.0	6.9	5.3	94.7	8.0

B Tris buffer

1 Preparation of 0.2 M stock solutions:

(a) Stock A: add 24.2 g of Tris(hydroxymethyl)aminomethane to 1 litre of water.

(b) Stock B: 0.2 M HCl.

2 Mix 50 ml of stock A with x ml of stock B and dilute to a total of 200 ml for the following pH of buffer:

x	pH
16.5	8.4
12.2	8.6
8.1	8.8
5.0	9.0

3 Purification strategy

3.1 Ordering of steps

The following section provides an overview of techniques used in protein purification. There is now a bewildering range of available techniques and separations products to choose from such that there are potentially many pitfalls awaiting those new to the science. The key requirements of the majority of purifications are:

(a) Release of product from source material—making an extract.

(b) Removal of solids—separation of cells and cell debris from soluble protein.

(c) Removal of water—concentration to simplify subsequent handling.

(d) Removal of contaminants—to be degree necessitated by purpose of purification.

(e) Stabilization of the product—to a degree dependent on lability and time before designated use.

As a general rule, initial steps in protein purification aim to generate a crude extract of the target protein along with other contaminants in a supernatant which is free of particulate matter. It is clearly important to start with a quantity of material (whether a volume of fermentation broth or a sample of animal tissue) which will yield the desired quantity of target protein following purification. Normally this is then followed by a number of adsorption and chromatography steps to generate the target protein at the required degree of purity. Chromatographic media vary in their capacity to bind protein and their cost. In general, high capacity, low cost media such as ion exchangers and group affinity media are used for initial purification of a protein from gross contaminants. Low capacity, high cost media such as affinity and HPLC matrices are used for a final step purification where a small volume column can be used to selectively remove any contaminants which remain. If the product is valuable (as with many therapeutic proteins) then yield following each purification step is most important. Speed of processing is often also important (even at the expense of resolution from contaminants) since it will limit the extent of protein degradation.

3.2 Nature of starting material

You may have little choice in the nature of the starting material for a purification. A fermentation which has previously been optimized to produce a high titre of a particular enzyme is a classic example. However it is advisable, where appropriate, to review the options for the starting material, and if following a recipe from a scientific paper, not to assume that the material used in the reference is necessarily the best to use for the job in hand.

The starting material for a protein purification may come from a wide variety of sources—animal tissue, plant material, milk, blood, and fermentation culture being typical examples. It is always important to maximize the total amount of product in the starting material and, if possible, minimize the amount of contaminants present. In this way the purification of the target protein will be simplified. For non-fermentation-based products this is often difficult since to a large extent one has to accept the starting material one is given. It is important to calculate that the starting material has a sufficient amount of the target product for the purification to be completed successfully.

Proteins are the products of ribosomal activity and are therefore associated with cellular metabolism. Consequently, regardless of the source, proteins are located either within or form part of the structure of, a cell or are secreted from a cell. Cellular-located proteins include insoluble inclusion bodies, soluble enzymes, or structural proteins which form part of a cellular structure such as a membrane. Proteins which are secreted from cells or localized in the periplasmic space may also provide simplified purification since the level of other protein contaminants is reduced compared to intracellular proteins. In addition, extracellular proteins (such as lysozyme and amylase) are generally more robust since they have to operate in a harsh environment. Intracellular proteins are generally less stable and present a greater challenge to the purification scientist due to their increased lability and presence of other proteins as contaminants. Where proteins are associated with cellular material it is advisable to generate a clarified (i.e. non-particulate) extract of the starting material as soon as possible through cell disruption, solubilization of the target protein in an appropriate buffer and removal of cells and cell debris through filtration or centrifugation. Where the protein is extracellular to cellular material, an initial separation to remove cells and leave a clarified supernatant containing the target protein is required. Thereafter the clarified preparation may be used immediately for purification or frozen in aliquots for future use.

3.3 Storage of material

Freshness of starting material is important since natural degradation processes and contamination from micro-organisms take place rapidly and result in a reduced level of the target protein and a non-representative starting material. Therefore unless purifications can be started within hours of delivery of the starting material, degradation should be slowed through reducing the temperature, to 5°C for a period of a few hours, or through freezing to −70°C for periods of days. Freezing can use the initial material or the product of an initial extraction. During the freezing process certain changes take place in the starting material which, if not checked, may reduce subsequent purification efficiency. Any starting material will probably have a high water content. As this water freezes ice crystals will grow which will disrupt cellular organelles and membranes. Secondly, salts will crystallize out after the water, with the less soluble salt in a pair coming out of solution first, causing a shift in the pH. Finally, proteases which are released from disrupted cells can cause a slow but irreversible degradation of a target protein.

To counteract these degradative processes and ensure that the target protein is preserved until a purification can be started it is advisable to freeze as soon as possible, preferably to a temperature of $-70\,^\circ$C so as to minimize any deterioration. In addition, protease inhibitors may be added to an extract and the pH stabilized through addition of a buffer. Where appropriate the starting material, or an initial extract, may be split into identical aliquots and frozen separately for subsequent repeat purifications. On recovery of the material from freezing rapid thawing should be used, with immersion of the material in warm water where possible.

3.4 Fermentation products

In fermentation there is considerable scope for increasing product titre and reducing contaminants through strain selection, genetic manipulation, and fermentation optimization. Contaminants may be derived from the ingredients used in the fermentation, added as it proceeds, or produced as a result of the fermentation along with the target protein. In addition certain contaminants such as albumin may protect the protein product and minimize degradation. Recombinant organisms such as E. coli strains frequently produce proteins as insoluble inclusion bodies which will require solubilization and refolding during the purification. Fermentations which use defined nutrients (as opposed to crude ingredients such as corn steep liquor and malt extract) may produce a feedstock which contains fewer contaminants, while minimizing antifoam addition may also reduce fouling problems in purification caused by antifoam blinding of membranes or adsorbents. Attention should be given to checking the optimum time for fermentation harvesting (often in log phase) and such evaluations should take into consideration the level of contaminants which may be problematical during purification as well as the concentration of the target protein. Clearly, this approach may require an interactive process in which initial purification work will identify the most problematical contaminants present in the fermentation.

3.5 Other sources of proteins

Other sources of proteins will include animal tissue such as muscle, heart, brain, or liver, plant tissue, blood, and milk. Animal tissue should be as fresh as possible and where possible it is recommended to use the local abattoir. Common starting materials include rat liver, rabbit muscle, beef and pig heart, liver, and kidney. Skeletal muscle should be allowed to go into rigor mortis so that ATP is reduced and actin and myosin do not partially solubilize during formation of the extract. Providing it is fresh, plant material can be obtained from the local shop. Yeast is a readily available source of proteins in fresh or dried form. Fresh baker's yeast contains active enzymes and should be frozen or chilled until use. Drying of yeast may inactivate some labile enzymes.

3.6 Making an extract

Many proteins such as enzymes are intracellular and necessitate preparation of an extract prior to further processing. Extraction technique should preferentially release the required protein, leave as many contaminants behind as possible, and minimize degradation of the protein product while further protein is being released. Extraction medium conditions must therefore be selected in which the target protein is stable and efficiently extracted. It should be noted that the pH for stability is not always the same as the pH which ensures maximum activity. Initial preparation steps should be carried out as fast as possible and, if the protein is particularly labile, at a low temperature. Many intracellular enzymes become unstable when removed from the stabilizing internal environment of the cell. Freezing (ideally to $-70\,^{\circ}C$) of cellular material in batch sizes appropriate to the purification scale provides a very convenient source of a consistent starting material.

A variety of disruption techniques exist, the most popular being freeze-thawing, bead milling, and homogenization. The various options for disruption permit selection of a technique appropriate to the nature of the cellular material. The disruption technique selected should never be more vigorous than required to lyse the cells since over-disruption may cause inactivation of enzymes once released. When isolating organelles such as mitochondria and chloroplasts the use of gentle disruption technique such as cell wall lysis with proteolytic enzymes may be evaluated to allow effective release of organelles without damage. However such lysis techniques may result in reduced yields such that a preferable option is to use a harsh disruption technique which while causing release of all cell contents, provides higher yields of target protein as a result. Sufficient extraction buffer (at least two volumes is recommended) must be used to ensure release of the remaining protein still trapped with the lysed cell material so as to ensure efficient recovery. At this stage it is wise to evaluate the optimum extraction time since prolonged exposure to the extraction method can lead to reduced product yields through denaturation. In addition it may be necessary to cool the cell slurry when using certain techniques such as homogenization or bead milling since heat is generated during the extraction process. Cells contain salts and other charged materials such as proteins and nucleic acids, giving an intracellular ionic strength typically between 0.15–0.2 M. When the intracellular contents are released into a lower ionic strength extraction buffer, charged particulates can act as ion exchangers and adsorb the target protein, so reducing yields. A buffer of ionic strength and pH similar to the intracellular conditions should therefore be used, with the addition of stabilizing agents to minimize denaturation. As an example I recommend 20–50 mM phosphate pH 7–7.5, with optional addition of EDTA, 2-mercaptoethanol, cysteine, and metal ions such as Zn^{2+}.

Plant material only contains a small volume of cytoplasm retaining intracellular enzymes since the majority of the intracellular volume is occupied by the vacuole. Consequently much liquid is released on lysing plant cells and although additional extracting liquid is not usually needed it is essential to control the pH and minimize protein inactivation since acidification and oxidation are likely after lysis. In addition phenolics are released which oxidize to form dark pigments which may interfere with the extraction process by attaching to proteins and inactivating enzymes. Thiol compounds such as 2-mercaptoethanol minimize the action of phenol oxidases and powdered polyvinylpyrrolidone may be used to adsorb the coloured phenolics during the extraction process. A typical protocol is to use 0.5–1.0 volumes of cold extracting buffer containing 20–30 mM 2-mercaptoethanol prior to disruption using a Waring blender for 30 sec followed by centrifugation. Polyvinylpyrrolidone may be added if needed.

Animal tissue is typically diced and trimmed to remove unwanted fat before shredding in a Waring blender for 30 sec with 2–3 volumes of extraction buffer per gram of tissue. The pH should then be checked before centrifugation at 5000–10 000 g for up to 60 min. The lysate should then be coarse filtered to remove fats in suspension.

Bacteria may be disrupted using ultrasonication, bead milling, or homogenization using a Manton-Gaulin homogenizer. Gram positive strains will easily disrupt in lysozyme, typically using 0.2 mg/ml egg white lysozyme at 37 °C for 15 min. Deoxyribonuclease I may be added at a concentration of 10 µg/ml to reduce the viscosity caused by DNA release. Gram negative strains may be lysed using a combination of detergent, osmotic shock, and lysozyme. After extraction the pH should always be checked and the suspension centrifuged to remove particulates. Cell debris will settle with spinning at 10 000 g for 15 min. Any remaining cloudiness will usually be removed during subsequent purification steps and fat globules may be removed using a coarse filter or glass wool. Inclusion bodies are commonly produced in bacteria with intracellularly expressed recombinant proteins and accumulate as insoluble aggregates which need to be solubilized and then refolded to recover the native state. At a laboratory scale, low speed fractional centrifugation allows for preferential separation of particles prior to treatment.

Yeast are typically disrupted using a Manton-Gaulin homogenizer or bead mill with use of two volumes of buffer per gram wet weight of yeast.

3.7 Separation of particulates

Purification techniques such as chromatography require that a protein sample is free of particulate material so as to prevent bed clogging, poor purification, build-up of back-pressure, and equipment break-down. Such particulate material is usually derived from cells, cells debris, and aggregated cell components such as proteins.

Consequently, particulate removal techniques such as centrifugation and filtration are commonly used following generation of an extract and before subsequent purification using chromatography. Centrifugation is the sedimentation of particles in an increased gravitational field. At a laboratory scale, batch centrifugation, in which samples are clarified in bottles or tubes in a rotor, is typically used. At a larger scale, continuous centrifugation is adopted to process the larger volumes. Differential centrifugation is also used to fractionate organelles. Filtration with filter papers can also be used, with suction applied using a Buchner funnel. This approach is less common since the filter papers tend to clog with fine particulate material and so reduce the flow of clarified supernatant. The flow-rate through the filter may be improved by adding filter aids such as celite with the material to be filtered prior to pouring into the Buchner funnel. Membrane processing using microfiltration and ultrafiltration are increasingly used, particularly to produce a more concentrated suspension of cells and cross-flow microfiltration is particularly useful for large scale purification.

3.8 Protein concentration

Following separation of particulates from the supernatant, a prime requirement is the removal of water to concentrate the target protein. This reduction in volume simplifies subsequent handling and reduces the time required for the purification steps to follow. Commonly used techniques include precipitation, ultrafiltration, adsorption, dialysis, and addition of dry gel permeation media.

Precipitation is commonly used early on in a purification strategy to concentrate and partially purify proteins. Following addition of ammonium sulfate the precipitate is separated from the supernatant (typically using batch centrifugation) and the precipitate redissolved in a smaller volume of an appropriate buffer. This technique is usually applied if the protein concentration is above 1 mg/ml, since lower concentrations may lead to denaturation or poor aggregation to form a precipitate.

Ultrafiltration using membranes with a 1000–300 000 Da cut-off separates low molecular weight molecules and water from proteins. In ultrafiltration, water and low molecular weight solutes are forced through a membrane to leave the target protein in a concentrated form in the retentate. Concentration through addition of dry gel filtration media is appropriate at a laboratory scale. Simple adsorption–desorption steps using a packed bed or batch adsorption achieve a similar result. Adsorption to concentrate usually uses ion exchange, with elution in a small volume using a stepwise increase in salt concentration.

Addition of a dry media of appropriate pore size will lead to the rapid uptake of water to swell the gel, with the exclusion of protein. The swollen media is then washed to remove protein from outside the particles prior to filtration under suction. Since the washing stage causes dilution, adequate concentration may only be achieved if the washing stage is avoided and this may lead to unacceptable yield losses.

Dialysis may be used to concentrate the protein sample and provide buffer exchange. For concentration purposes, water-attracting polymers such as carboxymethyl cellulose and polyethylene glycol are commonly used. The protein sample is placed inside a bag formed from a strip of dialysis tubing which is tied and immersed in a solution containing the polymer which will then attract water from the bag and concentrate the protein sample.

3.9 Adjusting sample composition between steps

Optimum integration of the processing steps used in protein purification will lead to a minimum need to adjust sample composition between each technique.

However it is rare that an entire purification will be carried out without a need for sample adjustment, typically involving alteration of pH and/or ionic strength. Such adjustment is necessary so that the protein sample has the required properties for the next purification step to work satisfactorily. Sample composition should always be checked prior to the next step and diafiltration, dialysis, and gel filtration can all be used to change the sample buffer. Wrong pH can be adjusted simply by adding acid or alkali while an excessively high ionic strength can be remedied through dilution. An appearance of turbidity is a common problem during purification, particularly as protein concentration increases, and centrifugation or filtration may be useful to reduce such haziness before continuing with a purification. However, the cause of such turbidity should be examined in some detail since although it may be easily removed at a small scale it may be more noticeable, and cause more serious problems as a process is scaled-up.

Dialysis is typically used for the removal of unwanted low molecular weight solutes from a sample and for their replacement with the buffer in the dialysate. A high concentration of salt in the dialysis bag causes water to enter while generally molecules larger than 15 000–20 000 Da do not pass out of the bag from the interior. Eventually, the buffer composition on each side of the dialysis membrane will equalize. It should be pointed out that dialysis does not remove unwanted solutes but just dilutes them with the dialysis buffer since the concentration reaches an equilibrium on either side of the dialysis membrane. Stirring the dialysis buffer and bag will help the process to speed-up reaching equilibrium which typically reaches 90% completion in 2–3 h. Dialysis is a convenient step to carry out overnight if no loss of protein activity is likely and the tubing is normally used from a roll, wetted, and a knot tied in one end. The protein sample is poured through a filter funnel into the open end of the tubing which is then knotted to seal the tube and placed in a beaker of the dialysis buffer. It is wise to leave an air space in the tube to allow for entry of dialysis buffer and expansion of the internal liquid volume.

Although dialysis is very simple and can be used overnight, since it can be a slow process, gel filtration is often used in preference to perform a buffer exchange. In gel filtration all the previous buffer is removed in a single quick step using a sample volume of less than 1/5 of the bed volume. The bed must be equilibrated in the desired buffer and the protein concentration kept to less than 30 mg/ml.

4 Protein lability and structure—implications for purification

4.1 Introduction

Proteins, and in particular, enzymes, are often sensitive molecules and the working environment of a protein is often indicative of the robustness of the molecule itself. As an example, intracellular enzymes are protected in a natural environment which is highly stabilizing—high in protein concentration, low in oxygen tension, and with reducing compounds present. As a consequence enzymes which are located within cells are often vulnerable to denaturation or loss of activity when exposed to the harsh extracellular environment following cell lysis. In contrast, extracellular enzymes are more robust since they are designed to work in an environment which may be highly oxygenated, exposed to proteolytic enzymes, and in which the protein concentration may be low.

During purification the target protein is taken from its natural working environment and exposed to conditions which, if not carefully controlled, may cause a loss in yield and a resultant poor purification efficiency.

There are three primary causes of loss of enzyme activity—denaturation, inactivation, and proteolysis.

4.2 Denaturation

This is caused by extremes of pH, temperature, and denaturants such as organic solvents. The interior of a cell commonly has a pH of between 6–8 and so it is normal practice to use buffers of similar pH during purification. When working outside this pH range, for example, during chromatographic separations, the stability of the protein should be closely monitored.

4.3 Catalytic site inactivation

This is a common cause of loss of enzyme activity. If inactivation is due to loss of cofactors, then simple addition will often restore enzyme activity. However inactivation may also be the result of covalent modification of the active site. Here cysteine residues are particularly prone to oxidation, leading to disulfide bond formation, partial oxidation to sulfinic acid, or irreversible oxidation to sulfonic acid. Disulfide bond formation is accelerated by the presence of divalent ions such as calcium. Consequently a common means of prevention of covalent modification is through the removal of heavy metal ions by addition of chelating agents such as EDTA at 10–25 mM. In addition, a sulfydryl-containing reagent such as 2-mercaptoethanol (5–20 mM) or dithiothreitol may be added. Alternative additives include glutathione thioglycolate and EGTA, which is more specific for calcium. When adding EDTA to a buffer, final pH adjustment should be carried out after addition of the chelating agent. It should also be noted that some proteins are actually stabilized by the presence of ions such as calcium and magnesium such that addition of metal ions may be necessary. Very dilute enzymes (e.g. 1 µg/ml) lose activity quickly and so it is recommended to keep the protein concentration high or add a stabilizing protein such as BSA (up to 10 mg/ml).

4.4 Proteolysis

This is commonly caused by the exposure of proteins to the digestive enzymes within cells. This often occurs as a result of cell lysis in which intracellular proteins which are normally separated through compartmentalization are rapidly mixed. Proteases are enzymes which cause the digestion of other proteins through hydrolysis and play an important role in cellular digestive processes. They are found in lysosomes in mammalian cells, in vacuoles in plants, and between the plasma membrane and cell wall in micro-organisms. Protein purifications which involve a cell lysis step in order to isolate intracellular enzymes will therefore potentially expose the target protein to such degradative enzymes, and steps may be necessary to minimize the activity of proteases so as to minimize yield losses. A common approach is to add chemicals to the initial extract which will slow or prevent the hydrolytic action of proteases. PMSF is the most commonly used additive, inhibiting serine, thiol, and some carboxypeptidase protease activity using a final concentration of 0.5–1.0 mM. In addition, pepstatin may be used to inhibit acid proteases. As a general rule, if protease activity is causing yield losses of the target protein, it is preferable to work as quickly as possible and, if possible, to carry out the purification at reduced temperatures (e.g. 5 °C).

A common approach to minimizing loss of activity is to use a standard cocktails of reagents which can be stored until used and added to an extract to ensure protein inactivation is slowed. A good combination is 2–5 mM EDTA, 0.5–1 mM PMSF, and 10^{-7} M pepstatin A.

4.5 Other causes of yield loss

In addition to the need to control pH and temperature, certain other precautions may be needed to prevent loss of target protein. These include minimizing bacterial growth, reducing water activity, and prevention of aggregation. Protein extracts which are produced during a purification, and buffers which are used during the process, are potentially suitable substrates for microbial growth, causing a hydrolysis of proteins through action of secreted enzymes and consequential loss of protein yields, in addition to potential microbial fouling of purification equipment. Consequently all extracts should be stored at 5 °C overnight, or frozen if left for longer periods. Similarly, all buffers should be freshly prepared and sterile filtered if used for chromatography.

Maintenance of a low water activity in protein preparations may also help to minimize yield losses. Glycerol, which forms strong hydrogen bonds to slow down the motion of water in buffers, may be used at up to 50% (w/v), although the increase in viscosity which results usually means that a 20–30% (w/v) solution is preferred. Sucrose and glucose are also used for a similar purpose. Any reagent added to protein extracts or buffers must be compatible with the purification techniques being used. Ammonium sulfate, for example, may be used to reduce water activity and is compatible with hydrophobic interaction chromatography, but is not suitable for addition to extracts prior to ion exchange since it will increase the salt concentration and prevent binding.

Certain proteins may be bound to insoluble cellular components such as membranes, or aggregated into insoluble particles, such as inclusion bodies. In these cases it is necessary to reduce hydrophobic interaction so as to solubilize the target protein and addition of detergents or chaotropic agents is common. Mild non-denaturing detergents such as Triton X-100 are preferred to strong detergents such as SDS which will interfere with protein structure and cause denaturation. Detergent levels added to protein preparations must also be kept below the critical micelle concentration. Chaotropic agents such as urea and guanidine hydrochloride are also useful, particularly for solubilization of inclusion bodies. Reducing agents are useful for use with bacterial enzymes derived from a reducing environment while mammalian cell enzymes are often best stabilized with surfactants and protease inhibitors.

References

1. Stoll, V. E. and Blanchard, J. S. (1990). In *Methods in enzymology* (ed. M. P. Deutscher), Vol. 182, pp. 34–5. Academic Press.

Chapter 3
Analysis of purity

Dev Baines

Prometic Biosciences, Unit 211, Cambridge Science Park, Milton Road, Cambridge CB4 0ZA, UK.

1 Introduction

1.1 Considering yield and purity

In approaches to protein purification, the concepts of yield and purity are routinely used, but are often difficult to define in absolute terms (1). To some extent the *purity* of a protein sample will be defined by the final designated use of the product. In most cases, analyses will involve measurement of the mass of the protein in the sample and quantitation of specific property of the target molecule (e.g. activity) to provide values for the *yield*. Thus, the calculation of specific activity of given fractions through the purification process provides a valuable indication of the level of the purity attained. Although, conventionally, specific activity is used in the purification of enzymes, due to the availability of sensitive and specific assays, recent developments in fast chromatographic separations and protein mass spectrometry have led to application of these techniques to address the purity of a wider variety of biomolecules.

The level of purity for any protein product requires several factors to be taken into consideration. Besides the intended use, the source of the protein will dictate the extent of analyses required, since the level of impurities present in the final product will depend not only on the purification process used but also on the starting source material. For bulk enzyme preparations (for use e.g. in biotransformations or related applications) it may only be necessary to ensure the product is free of any contaminating activities which could effect the outcome of these types of applications.

For proteins required for physical studies (protein crystallography, primary sequence analysis), the purity criteria are more stringent, particularly, since the lack of purity (including sample microheterogeneity) can drastically influence the outcome of such studies. Alternatively, proteins intended for therapeutic use will have purity considerations significantly different, constrained not only by regulatory requirements but also by clinical responses that may arise from the presence of any contaminants. The nature of these contaminants, as mentioned earlier, will depend on the starting source of the target molecule (i.e. animal tissue, human serum, recombinant micro-organisms [prokaryotes, eukaryotes], and hybridomas). Of concern is the antigenicity of not only the source derived contaminants but also structurally modified forms of the product which may be more immunogenic than the authentic forms. Contamination by the source DNA, agents capable of transmissible diseases (viruses, prions), and presence of pyrogenic components requires that therapeutic proteins should be shown to be pure by not only the conventional techniques, but also by additional methods for measuring contaminants which are of clinical relevance (2).

While several methods are available for the analysis of *purity* or more strictly the impurities in the sample, the choice of tests will, however, depend on the amount of protein available, the nature of impurity being tested, the accuracy of the estimate, the sensitivity of the assay, and any incompatible properties of the protein and the solvent in which the protein is dissolved (3).

Consequently, any demonstration of protein purity always requires an assessment of types of the impurities present in the final product. Thus, it is important to identify the type of purity that is to be measured and then provide an identity test for it to allow its distinction from the target protein. By definition, *purity* is, therefore, a demonstration that the sample is free from detectable quantities of contaminants.

1.2 Scope of the methods included

In this chapter, brief details of generic methods used for the analysis of *purity* are given, which allow for routine analysis of fractions through the protein purification process and are readily available in most laboratories. A brief mention of emerging tools which complement the traditional protein purity analysis is given. All methods are applicable to proteins produced from diverse sources, although some appropriate modifications may be necessary. Since protein purification strategies are designed to enrich the protein component of interest from complex starting materials, this chapter will focus on methods for estimating protein contaminations of the target product. General assays for the presence of other potential contaminants can be obtained from the literature (3–5).

2 Total protein quantitation

2.1 Ultraviolet absorption protein assay

Proteins show maximal ultraviolet (UV) absorption at about 280 nm arising from the contribution made by aromatic residues (of tryptophan, tyrosine) and to some extent from cystine groups. Absorption at 280 nm can, therefore, be used to provide an approximate protein assay since, the intensity of absorption bears some relationship to the number of these residues present in the protein. However, an accurate absorption spectrum cannot be predicted by simply summing up the absorption of the number of these residues in a given molecule. This is because the observed absorption spectrum reflects the varying environment of amino acid residues within the protein and the solvent (tyrosine side chain contains an acid–base function; the absorption spectrum of the undissociated phenolic form [neutral pH] is significantly different from the dissociated phenolate form [alkaline pH]). Nevertheless, provided the amino acid sequence is known, a close approximation of the absorbance coefficient of a protein at 280 nm can be calculated by the following protocol based on the average values of molar absorbance coefficient for the chromophores present as described in companion volume to this series (6).

Protocol 1

Calculation of the absorbance coefficient of protein

1   Count the number of tryptophan residues (η_{Trp}) and tyrosine residues (η_{Tyr}) in the sequence.

2   Count the number of disulfide bonds (η_{ss}) and if unknown assume that for intra-cellular located proteins the value is zero and for secreted proteins $\eta_{ss} = \eta_{Cys}/2$.

3   Use the following equation to calculate absorbance coefficient:

$$\varepsilon_{280} \,[\text{M}^{-1}\text{cm}^{-1}] = 5500 \times \eta_{Trp} + 1490 \times \eta_{Tyr} + 125 \times \eta_{ss}$$

Note that absorbance measurements at 280 nm is subject to interference from the presence of other compounds that have absorbencies around this wavelength, including nucleic acids. The estimation of the protein concentration in the presence of nucleic acids is possible by measuring absorbencies at 280 and 260 nm wavelengths and applying the following correction factor (3, 4):

$$\text{Protein (mg/ml)} = 1.55\, A_{280\,\text{nm}} - 0.76\, A_{260\,\text{nm}}$$

For many known proteins, molar extinction coefficients may be available from reference sources (3).

Protocol 2

Measurement of protein concentration by UV spectroscopy

Equipment and reagents

- UV spectrophotometer
- Protein solution
- Buffer blank

Method

1   Switch on the UV spectrophotometer and set wavelength to 280 nm. Leave the instrument to stabilize (ca. 15–30 min).

2   Zero the instrument with cuvettes containing the buffer in which the protein is dissolved.

Protocol 2 continued

3   Measure the absorbance of the protein solution by replacing the buffer blank with the cuvette containing the test sample.

4   Protein concentration can then be estimated using the following conversions when using a cuvette with a 1 cm path length:

$$\text{concentration (mg/ml)} = \text{absorbance}/A^{1\ \text{mg/ml}}$$

$$\text{concentration (M)} = \text{absorbance}/\varepsilon_{\text{m}}$$

where ε_{m} is the molar extinction coefficient.

This assay requires the solution to be at ambient temperature (to avoid atmospheric moisture condensation onto the cuvette surface) and to be free of any presence of particulate matter.

2.2 Lowry (Folin-Ciocalteau) protein assay

This is the most widely-used method for quantitative determination of protein concentration. Reaction of the phenolic moiety of tyrosine in protein with Folin-Ciocalteau reagent, which contains phosphomolybdic/tungstic acid mixture produces a blue/purple colour with absorption maximum around 660 nm. Additionally, the use of a copper reagent enhances the colour formation by chelating with the peptide bonds and allowing for efficient electron transfer to the chromophore formed. This method is sensitive down to 10 μg protein per ml. The Folin-Ciocalteau reagent is commercially available.

Protocol 3

Protein estimation by Lowry method[a]

Equipment and reagents

- Spectrophotometer
- 2% sodium carbonate in 0.1 M sodium hydroxide
- Copper reagent
- Folin-Ciocalteau reagent

Method

1   Prepare 2% (w/v) sodium carbonate in 0.1 M sodium hydroxide—solution A.

2   Prepare 1% (w/v) copper sulfate—solution B.

3   Prepare 2% (w/v) sodium potassium tartrate—solution C.

4   Prepare the copper reagent—solution D. Mix 0.5 volume of solution B, 0.5 volume of solution C, and 50 volumes of solution A.

5   Folin-Ciocalteau reagent is diluted to 1 M acid according to the suppliers instructions—solution E.

Protocol 3 continued

6   To a 1 ml protein solution (20–200 μg protein) add 5 ml of solution D, mix thoroughly by vortexing, and stand at room temperature for 10 min.

7   Add 0.5 ml of solution E, mix rapidly, and incubate for 30 min at room temperature.

8   Measure the absorbance at 750 nm against reagent blank not containing protein. The concentration is estimated by referring to a standard curve obtained at the same time using known concentrations of bovine serum albumin.

[a] Note that many of the commonly used reagents, e.g. Tris, Pipes, Hepes, EDTA, and detergents are known to interfere with this assay.

2.3 Bradford (Coomassie Brilliant Blue) dye-binding protein assay

Coomassie Brilliant Blue complexes with proteins to give an absorption maximum at 595 nm. It is a simple method with the colour developing rapidly to produce a stable complex and is sensitive down to 20 μg protein per ml, but the amount of dye binding to the different protein molecules is variable and does require a careful selection of a protein standard for generating the calibration curve.

Although dye-binding reagent (Bradford) is easily prepared in the laboratory, it is more convenient to purchase it in pre-prepared form. Suppliers include Pierce, Bio-Rad, and Sigma.

Protocol 4

Protein estimation by dye-binding assay

Equipment and reagents

- Spectrophotometer
- Coomassie Brilliant Blue
- Phosphoric acid
- 1 M sodium hydroxide

Method

1   Either use commercially available reagent or prepare by dissolving 100 mg of Coomassie Brilliant Blue G-250 in 50 ml ethanol. To this add 100 ml of phosphoric acid (85%, w/v) and dilute to 1 litre with water.

2   Switch on the spectrophotometer and allow to warm up for at least 15 min.

3   Accurately pipette 20 μl samples into 1 ml disposable polystyrene cuvettes and add 50 μl of 1 M NaOH to the sample. (Note that the commercially available reagents do not require this addition of sodium hydroxide solution.)

4   Add 1 ml of the dye reagent, mix, and incubate for 15 min at room temperature.

5   Measure the absorbance at 595 nm. The concentration of the sample is obtained from a standard curve obtained by using known concentration of standard protein.

Note: use of a disposable polystyrene cuvettes is recommended as the dye tends to stick to the cuvette.

2.4 Bicinchoninic acid protein assay

Bicinchoninic acid (BCA) reagent provides a more convenient and reproducible cuprous ion base assay and has a sensitivity similar to the Lowry method, although it is subject to interference from DTT concentrations above 1 mM (7). The sensitivity range for this assay is 0.2–50 μg. The reagents may be obtained from commercial suppliers or prepared as described in *Protocol 5*.

Protocol 5

Protein estimation by BCA method

Equipment and reagents

- Volumetric flask
- Spectrophotometer
- Disposable polystyrene cuvettes
- Sodium bicinchoninate
- Na_2CO_3

- Sodium tartrate
- NaOH
- $NaHCO_3$
- $CuSO_4 5H_2O$

A Reagent preparation

1 Dissolve 1 g sodium bicinchoninate, 2 g Na_2CO_3, 16 g sodium tartrate, 0.4 g NaOH, and 0.95 g $NaHCO_3$ in 90 ml of water. Adjust the pH to 11.3 with 10 M NaOH and make up to 100 ml in volumetric flask (reagent A).

2 Dissolve 0.4 g $CuSO_4 5H_2O$ in 10 ml of deionized water (reagent B).

3 Freshly prepare working solution by mixing 50 volumes of reagent A with 1 volume of reagent B.

B Method

1 Add 1 ml of the working reagent to a 24 μl sample and mix thoroughly. Incubate at 60°C for 30 min. Cool the samples to room temperature. The samples are stable up to 1 h following incubation at 60°C.

2 Read the absorbance at 562 nm using disposable polystyrene cuvettes.

3 The protein concentration in the samples is estimated using a standard curve (0–50 μg standard protein) obtained at the time of the assay.

Note: for greater sensitivity increase the incubation time, and for lower sensitivity decrease the incubation temperature to ambient.

2.5 Colloidal gold protein assay

Metal-binding assays are, particularly, useful for measuring low concentrations (20–640 ng) of proteins. The colloidal gold protein assay is shown in *Protocol 6*.

Commercially formulated colloidal gold protein assay reagent used in the original description of the assay has been changed and cannot be used for the protein assay. However, the author has described the method in detail (8).

Protocol 6

Estimation of protein by colloidal gold protein assay

Equipment and reagents

- 250 ml flask
- Magnetic stirrer bar
- Spectrophotometer
- Microcentrifuge tubes
- Chloroauric acid
- Trisodium citrate dihydrate
- Tween 20
- Citric acid

A Reagent preparation

1 Using clean glassware, gently bring 80 ml of distilled water to boil in a 250 ml flask containing a magnetic stirrer bar.

2 To this add 100 μl of 40 mg/ml chloroauric acid with stirring.

3 Add 1 ml of 40 mg/ml trisodium citrate dihydrate after the gold has dissolved, and boil the solution for 30 min with refluxing.

4 Cool the colloidal suspension to room temperature add 32 μl of 25% Tween 20, mix thoroughly, and add 400 μl of 1 M citric acid to the solution.

5 Store the clear red solution at 4°C.

B Assay procedure

1 Pipette 1 ml of the gold reagent into 1 ml microcentrifuge tubes.

2 Immediately before use, dilute protein standard into 0.01% Tween 20.

3 Add 10 μl of diluted protein standard solution and the unknown samples into the gold solution tubes. Mix thoroughly by using a vortex mixer and leave at room temperature for 15 min.

4 Read the absorbance at 560 nm. Use protein blank to zero the spectrophotometer.

5 Obtain the protein concentration from a graph of absorbance plotted against standard protein solutions.

Note that the use of dirty or contaminated glassware in the preparation of the reagent will yield cloudy solution which cannot be used. Do not freeze diluted protein standards or store for prolonged periods to minimize protein adsorption to the tubes.

2.6 General comments

Since different proteins react to different extents with various colorimetric assay reagents, the absolute value for the target protein would be subject to these inherent limitations in methods. Bovine serum albumin is often most used as a protein standard in most protein quantification methods to obtain a standard curve. The purity of this standard (additives, moisture) will affect the values obtained from the standard curve. Prepared standards are available from most suppliers of reagents and should be used where possible.

If sufficient quantity of the target protein is available, then the same protein should be used to generate the standard graph. The protein should be dialysed against distilled water and freeze–dried. The dried powder should be stored desiccated at −20°C. Stock solutions (1–2 mg/ml) in water and/or appropriate buffer can be prepared and stored in working portions at −20°C.

3 Specific quantitation

3.1 Activity assays

Determination of the activity of a protein is based on the change in specific property of either the protein itself or an added component under controlled conditions. These methods can either be continuous (kinetic analysis), coupled (indirect kinetic analysis), or discontinuous (fixed time sampling). The first two approaches do not require a separation of the reactants to measure changes which occur on addition of the test sample. The third approach requires termination of the reaction in the incubation mixture and subsequent measurement of the changed component.

From thermodynamic considerations, it is important that the transformed component is used in excess with suitable controls which are the same as the test conditions, except for addition of either the substrate or the enzyme. Changes monitored in the measured parameter in the test as compared to the control experiment provide an indication of the extent of activity due to the sample. To check for the presence or the absence of activators/inhibitors and other assay artefacts, it is useful to assay the material using different sample volumes.

3.1.1 Enzyme activity measurements

Many substrates and reaction products have an absorbance in the visible or the UV region of the spectrum. It is not surprising, therefore, that a large number of enzyme assays are based on measuring the change in the UV/visible region of the spectrum as a consequence of the enzyme-mediated response. The rate of reaction is a measure of the amount of substrate converted or the product formed and quantifies the reaction in a unit time under specified conditions. Thus, the rate of an enzyme-catalysed reaction is the property of the enzyme molecule, although the observed rate can be influenced by the reaction conditions.

To monitor purity by enzyme activity measurements, the use of steady-state kinetics (typical reaction half-lives of greater than 10 sec) is the preferred method and the continuous procedure is the most convenient. After mixing the reactants, the changes in the property of the substrate or product are monitored continuously. Changes in extinction coefficient, fluorescence, and hydrogen ion concentration can be measured most accurately when connected to a recording system to allow the reaction progress curve to be recorded.

In most circumstances, initial velocities are measured. It is important to control the temperature, pH (hydrogen ion concentration), and the ionic strength of the reaction mixture as these variants can affect the enzyme assay. Changes in temperature can affect the rate constant by increasing or decreasing the reaction according to the Arrhenius law and by inactivating the enzyme. The reactions should, therefore, be carried out in a thermostated reaction container and the temperature always specified. The pH of the reaction mixture should be in a buffered system compatible with the reacting molecules, and at a buffer concentration to allow for any changes in the hydrogen ion concentration that may occur during the time course of the reaction. The ionic strength of the reaction mixture affects the reactivity of the reactants and should be controlled and specified.

Spectrophotometric procedures are the methods of choice for a wide variety of enzyme activity measurements. Generally, three main types of chromophores are utilized in these types of assays; the naturally occurring cofactors, synthetic substrates, and coupled reactions.

NAD and NADP coenzymes undergo absorption changes during the course of many dehydrogenase catalysed reactions. The oxidized form (NAD, NADP) of the coenzymes show little absorbance at 340 nm whereas the reduced form (NADH, NADPH) display a significant absorbance peak at this wavelength. Reaction of the coenzyme is stoichiometric with the substrates and can be monitored continuously at this wavelength. Synthetic substrates which liberate a chromophore are useful for monitoring enzymes whose natural substrates/products do not contain a convenient chromophore. For example, many proteinases are easily monitored using p-nitrophenyl anilide analogues of amino acids. The substrates are not absorbing by themselves but on proteolytic cleavage, the p-nitrophenolate ion is released which has an intense absorbance at 410 nm. Several of such synthetic substrates are commercially available (Bachem, Sigma, Boehringer) and assays have been described in the literature. In the coupled assays, the reaction of an enzyme which does not allow the use of either the naturally occurring chromophores or synthetic substrates is coupled to a second reaction which does have the required absorbance characteristics.

3.1.2 Enzyme unit definition and specific activity

In a description of the enzymatic activity, the following key information is required:

(a) The amount of substrate transformed or the amount of product formed in a concentration unit (e.g. μmol).

(b) The time of the reaction (e.g. min).

(c) The amount of protein in the reaction mixture (e.g. mg).

The enzyme activity is quoted as μmol/min and this is described as a unit. Thus, 1 unit of enzyme activity is the amount of enzyme which catalyses the transformation of 1/μmol of substrate per minute. The amount of enzyme activity per unit of protein is obtained as units per mg protein and is the specific activity of a given preparation of material.

For purity analysis, therefore, the determination of the specific activity will provide an indication of the purification achieved after each step. An effective purification step should show a high increase in the specific activity value.

3.2 Binding assays

3.2.1 Immunological detection of proteins

Immunoassays exploit the binding specificity of an antibody for its specific antigen and for quantification—either the antigen or the antibody has to be labelled with a reporter molecule. Antibody specificity for proteins (strictly protein surface epitopes) provides a convenient means of identifying and quantifying the target molecule as well as providing a means of assaying for impurities. Production of antibodies for this application is beyond the scope of this chapter and the reader is referred to published reviews (9, 10) and more appropriate laboratory manuals.

Assuming that antibodies directed to both the target protein and the potential contaminants are available, several assays including Western blotting (Section 4), enzyme-linked immunosorbent assays (ELISA), and dot-blot assays can be developed to provide an indication of the purity during protein purification. For greater sensitivity, radioactive isotopes are used as labels, and radioimmunoassay provides measurements down to fmol amounts of antigen (11). More recently, the availability of biosensors (e.g. BiaCore) has extended the usability of immunoassays to provide real-time assays.

There are a number of ELISA methods available for detection of both the antigens and the antibodies. The basic principles, using a 96-well-plate format is outlined in *Protocol 7*.

Protocol 7

ELISA for the detection of the target protein (antigen)

Equipment and reagents

- Microtitre plates
- Microtitre plate reader
- Antigen coating buffer: 0.1 M $Na_2CO_3/NaHCO_3$ pH 9.6
- Assay buffer: 20 mM phosphate-buffered saline (PBS) pH 7.4 containing 0.05% Tween 20

- Blocking buffer: 20 mM PBS containing 2% non-fat dried milk
- Primary antibody
- Conjugated secondary antibodies directed towards the primary antibody
- Substrate

Protocol 7 continued

A. Antigen coating to microtitre plates

1 Dilute antigen standard and the unknown sample in a concentration range of 0.1–10 μg/ml in the antigen coating buffer, 0.1 M $Na_2CO_3/NaHCO_3$ pH 9.6.

2 Add 50 μl to each well. Control wells contain no antigen and standard wells contain serially diluted antigen to provide data to construct the calibration graph.

3 Incubate the plates at room temperature for 5–6 h on microtitre plate shaker. Wash the wells three times with assay buffer.

4 Add 100 μl of blocking buffer and incubate overnight to block the non-specific binding sites.

5 Thoroughly wash the plates again with the assay buffer.

B. Primary antibody binding step

1 Experimentally determine the dilution required for primary antibody (directed either to target protein or impurity) containing solution (polyclonal serum or purified monoclonal antibody) by diluting in the solution in blocking buffer. For most applications dilution in the range 1:200 to 1:1000 is sufficient.

2 Add 100 μl of diluted antibody to each well and incubate for 5–6 h at room temperature.

3 Wash the plates out with the assay as before.

C. Secondary antibody binding step

1 Conjugated secondary antibodies directed towards the primary antibody (e.g. rabbit anti-mouse or sheep anti-rabbit) are used in this step. These antibodies are conjugated to peroxidase or alkaline phosphatase. Alternatively, peroxidase-conjugated avidin or streptavidin is used if the antigen or the primary antibody has been biotinylated.

2 Add 100 μl of the diluted (in the assay buffer) secondary antibody to each well and incubate at room temperature for 1 h.

3 Thoroughly wash the plates out with assay buffer.

D. Detection step

1 Add suitable substrate solution to each well. For peroxidase system add 100 μl of 0.8 mg/ml o-phenylenediamine dihydrochloride dissolved in 0.1 M citrate buffer pH 5.0, containing 0.3% hydrogen peroxide and incubate at room temperature for 30 min.

2 Read the plates at 450 nm on a microtitre plate reader.

3 Estimate the (target or impurity) protein from a standard graph.

4 Detection of impurities

4.1 Sodium dodecyl sulfate–polyacrylamide gel electrophoresis

SDS–PAGE (*Protocol 8*) is an excellent method for rapidly assessing the purity of proteins, and is routinely used in the development and validation of a purification strategy. When this technique is combined with blotting methods, it provides an extremely powerful method for demonstrating the protein identity. It is the most widely used method for analysing protein samples in a qualitative manner and the method is based on the separation of proteins according to size. The method can therefore be used to determine the relative molecular mass of the proteins as well as provide an analysis of the purity. A number of systems have been developed and readers are advised to refer to refs 12 and 13 for further information on the practice and theory of electrophoresis.

Various vendors supply both the equipment and reagents. The electrophoresis units tend to vary in design, and many suppliers now have pre-cast gels as a part of their inventory. Additionally, an automated system capable of running mini electrophoresis gels is available from Pharmacia (The Phast System) but the resolution of complete samples may be better in the conventional gels of 20 cm long and 16 cm wide. Due to toxicity of acrylamide monomer (neurotoxin), it is advisable where possible to use the available pre-cast gels. Acrylamide gels are conventionally characterized by two values (%T and %C) where %T is the total weight percentage of the monomers (acrylamide + bisacrylamide cross-linker) and %C is the proportion of cross-linker as a percentage of the monomer.

Protocol 8

SDS–PAGE electrophoresis of proteins

Reagents

- Acrylamide concentrated stock (30%T, 2.7%C): prepare stock acrylamide solution by dissolving 29.3 g of acrylamide and 0.8 g of bisacrylamide in 70 ml of deionized water. Make the final volume to 100 ml, filter the solution through a 0.45 μm membrane, and store at 4°C for a maximum period of four weeks. To dispose of the acrylamide solution, polymerize as described below and discard the gel.

- Resolving gel buffer (1.5 M Tris–HCl): dissolve 18.2 g Tris base in 70 ml of deionized water, adjust the pH to 8.8 with HCl, and make to 100 ml with water; store at 4°C

- Stacking gel buffer (0.5 M Tris–HCl): dissolve 6.1 g Tris base in 70 ml deionized water, adjust pH to 6.8 with HCl, and make up to 100 ml; store at 4°C

- 10% SDS solution: dissolve 10 g of SDS in 70 ml of deionized water and make up to 100 ml; store at room temperature

- Sample buffer: mix 4.8 ml of deionized water, 1.2 ml of 0.5 M Tris–HCl pH 6.8, 2.0 ml of 10% SDS, 1.0 ml of glycerol, and 0.5 ml of a 0.5% solution of bromophenol blue (w/v in water). Store at room temperature. For reducing buffer, add 50 μl of 2-mercaptoethanol to 950 μl of the sample buffer.

- Polymerization catalysts: freshly prepare 10% ammonium persulfate solution in water. *N,N,N',N'*-tetramethyethylenediamine (TEMED) is used directly from the supplied bottle which should be stored in dark.

- Electrode buffer (5 ×): dissolve 15 g Tris base, 72 g glycine, and 5 g SDS in 1 litre of deionized water and store in a clean glass container; for use, dilute one part of this with four parts water and check that pH is near 8.3

A. Casting gels

1 Ensure that all the plates, spacers, combs, and buffer chambers are clean. Always wear gloves while assembling the equipment and casting the gels. The resolving gel is cast first and then overlaid with the stacking gel.

2 Set up the casting apparatus and determine the gel volume to provide sufficient height of the resolving gel ensuring that sufficient space (1–2 cm) below the comb is available for the stacking gel.

3 Prepare the solution for appropriate percentage resolving gel by mixing the solutions according to desired acrylamide concentration (%T) except for the catalyst solutions (ammonium persulfate and TEMED). Degas the mixed solution under vacuum for 15 min. The following solutions should be used for 10%T gels:

   - water   4.02 ml
   - resolving gel buffer   2.5 ml
   - 10% SDS   0.1 ml
   - stock acrylamide solution   3.33 ml

4 Add the catalyst solutions (50 μl of 10% ammonium persulfate and 5 μl TEMED) into the degassed monomer solution and mix gently. Transfer the solution between the plates to the mark for the height of the resolving gel. Overlay the solution with water saturated 2-butanol to exclude air and allow the gel to polymerize for 1 h. Allow the unused gel to polymerize before discarding. Other concentrations of the acrylamide are achieved by varying the proportion of water and the stock acrylamide solution in the mixture.

5 Mix the following solutions of the stacking gel components:

   - water   6.1 ml
   - 0.5 M Tris–HCl pH 6.8   2.5 ml
   - acrylamide solution   1.3 ml
   - 10% SDS solution   0.1 ml

6 Degas the solution under vacuum for at least 10 min. Rinse the top of the resolving gel with water and dry off the residual liquid with a clean paper towel. Add 50 μl of 10% ammonium persulfate and 10 μl of TEMED into the stacking gel solution and gently mix. Pour the stacking gel solution on top of the resolving gel, carefully insert and align the comb ensuring that no air bubbles are trapped under the teeth. Allow the gel polymerize for at least 1 h.

Protocol 8 continued

B. Sample preparation

1   Mix equal volumes of sample with sample buffer in a 1 ml microcentrifuge tube, mix, and heat in boiling water-bath for 5 min.

2   For native and non-reducing gels the sample is mixed with the sample buffer not containing SDS or mercaptoethanol and is not heated.

3   Allow to cool to ambient temperature.

C. Electrophoresis

1   Assemble the electrophoresis equipment according to the manufacturer's instructions and fill the electrode chambers with diluted electrode buffer.

2   Gently remove the comb from the stacking gel and load the samples into the wells using a micropipette. The amount of sample loaded will depend on the width of the wells and the thickness of the gel and it is possible to load up to 50 μl without compromising the resolution.

3   Immediately, attach the leads to power supply. The lower electrode chamber is connected to the anode and the upper one to cathode in SDS–PAGE. Run the gels under constant current conditions (approx, 20 mA/mm of gel thickness) until the bromophenol tracking dye reaches the bottom of the resolving gel.

4.2 Detection of protein bands in electrophoresis gels

Routinely, two methods are used for the visualization of proteins in the SDS–PAGE gels. For Coomassie Blue R-250 staining (*Protocol 9*) the detection limit is about 1 μg of protein per band. The silver staining (*Protocol 10*) is more sensitive, requiring 0.1 μg protein per band and is more often used to assess the purity of the product. Following electrophoresis, the gel is removed from the two plates, the stacking gel is cut off and discarded. The gel is then placed into a staining or fixative solution depending on the procedure used to visualize the protein bands. During staining procedures, always wear gloves to minimize staining due finger marks.

Protocol 9

Coomassie Brilliant Blue staining of SDS–PAGE gels

Equipment and reagents

- Magnetic stirrer
- Coomassie Blue
- Methanol, acetic acid, and water (40:10:50, by vol.)

Method

1   Dissolve 0.1% Coomassie Blue R-250 into methanol, acetic acid, water (40:10:50) using a magnetic stirrer. Filter solution through filter paper and store at room temperature.

Protocol 9 continued

2 Stain the electrophoresis gel by soaking in the staining solution with gentle agitation for 1 h.

3 Wash the gel in water to remove excess stain and destain using methanol, acetic acid, water (40:10:50) with several changes until the background is sufficiently clear.

Note that extensive destaining can lead to loss of some of the low molecular weight bands and fading of the other bands. For permanent records, the stained gels can be photographed or dried using commercially available gel dryers.

Protocol 10

Silver staining of SDS–PAGE gels

Reagents

- Methanol, acetic acid, and water (40:10:50, by vol.)
- Ethanol, acetic acid, and water (10:5:85, by vol.)
- Silver nitrate solution
- Oxidizing solution: 3.4 mM potassium dichromate and 3.2 mM nitric acid
- Developing solution: 0.28 mM sodium carbonate and 1.9% formaldehyde

Method

1 After electrophoresis, fix the gel in methanol, acetic acid, water (40:10:50) for 1 h with gentle agitation.

2 Wash the gel twice for 30 min in ethanol, acetic acid, water (10:5:85).

3 Immerse the gel in oxidizing solution for 10 min.

4 Wash the gel for 10 min in excess water. Repeat the process three to four times until the pale yellow colour is washed out.

5 Soak the gel for 30 min in freshly prepared 1.2 mM silver nitrate solution.

6 Wash the gel for 2 min in excess water.

7 Place the gel in developing solution for 1 min. Replace the developer solution and incubate for 5 min for protein bands to appear. Replace the developing solution and continue incubating until bands are stained satisfactorily.

Silver staining reagent kits are supplied by Bio-Rad and Novex. It is important to maintain a high degree of cleanliness during the electrophoresis and silver staining to minimize artefacts (e.g. bands in the region of 50–10 kDa are often due to contaminants introduced during the sample preparation step).

4.3 Analysis by isoelectric focusing (IEF)

Proteins carry positive, negative, or zero net charges depending on the pH of the solution. The pH at which the protein has net zero charge, the isoelectric pH, is termed the pI (the isoelectric point) and is a characteristic physicochemical property of each protein. Under electrophoresis, the net charge determines the direction that protein migrates to and thus, at pH below the pI, a particular protein will migrate towards cathode and above its pI, it will migrate towards anode. A protein at its pI will not migrate in either direction. In IEF (*Protocols 11–13*) this concept is used to separate proteins on the basis of differences in their isoelectric point as they migrate through a pH gradient.

IEF is carried out under native conditions and can resolve proteins differing in their pI by 0.02 pH units. The sensitivity of IEF to small differences in isoelectric point can provide a very highly sensitive method for identification of potential contaminants present in the product. In combination with other electrophoresis techniques (e.g. two-dimensional electrophoresis) this sensitivity can be greatly enhanced (14).

There are several ways of carrying out analytical IEF. The method presented in *Protocol 11* has been modified for use with commercially available (e.g. Bio-Rad) horizontal flat bed apparatus and polyacrylamide gel.

Protocol 11

Gel preparation

Reagents

- Acrylamide solution
- 25% glycerol
- FMN
- 40% ampholyte
- 10% persulfate
- TEMED

Method

1 Prepare acrylamide solution (25%T, 3%C) by dissolving 24.25 g acrylamide and 0.75 g bisacrylamide in 70 ml of water. Make the volume to 100 ml and filter through a 0.45 μm filter.

2 Prepare 25% (w/v) glycerol.

3 Dissolve 0.05 g of riboflavin 5′-phosphate (FMN) in 50 ml of water and store in a container in the dark at 4°C.

4 Freshly prepare 0.1 g ammonium persulfate in 1 ml of water.

5 Use a clean glass IEF plate and place a few drops of water on it. Place the hydrophobic side of the polyester support film against the plate and allow the water drops to spread evenly below the film. Ensure no air bubbles are trapped under the film and remove excess water from the edges of the plate. Assemble the glass plate on the casting tray with the gel support film facing down.

6   Mix the following solutions in a clean vacuum flask:
- water   6.6 ml
- acrylamide solution   2.4 ml
- 25% glycerol   2.4 ml
- 40% ampholyte (w/v)   0.6 ml

7   Degas the above solution under vacuum for 15 min.

8   Add the following polymerization catalyst to the flask and mix gently:
- 0.1% FMN   0.06 ml
- 10% persulfate   0.018 ml
- TEMED   0.004 ml

9   Using a pipette, introduce the solution between the support film and the casting tray, ensuring no air bubbles are trapped. Illuminate the assembly with a fluorescent lamp placed 4 cm above the gel for 1 h. Gently lift the gel from the tray, turn the plate over with the gel side up, and place it under the lamp for a further 30 min.

4.3.1 Sample preparation and application

IEF will tolerate a small volume load (10–15 μl) of typical biological buffers but it is recommended that for most applications, the samples should be exchanged into water or 2% ampholytes using dialysis or gel filtration. The simplest method is to use Sephadex G25 columns for this purpose (PD-10 pre-packed columns are available from Pharmacia).

Although many methods are available for applying the sample to the gel, for thin-layer polyacrylamide gels, the best method is to place a filter paper soaked with the sample solution directly onto the gel surface. For the size of the gel described here, application of the sample is best achieved using 0.2 × 1 cm paper loaded with 5 μl of the sample (5–10 μg protein for Coomassie Blue dye staining). The loaded paper can be applied to any part of the gel, but to protect the sample from extremes of pH, it is preferred that the sample is positioned on or near the middle of the gel.

Protocol 12

Electrofocusing

Equipment and reagents

- Electrolyte solution: 1 M phosphoric acid and 1 M NaOH
- IEF apparatus

Method

1   The IEF cell should be connected to a refrigerated circulating bath and cooled to 4°C.

Protocol 12 continued

2   Apply a few drops of water to the horizontal platform to provide a good contact with the gel and place the gel (supporting film side down) on to the platform ensuring an even spread of the water. Avoid trapping any air bubbles.

3   Wet the electrode strips with the appropriate electrolyte solution; for analytical purposes use 1 M phosphoric acid for the anode (acidic) and 1 M NaOH for the cathode (basic). Remove excess solution from the electrode strips using paper towels and place them along the appropriate edges of the gel. The electrode strips should not extend beyond the sides of the gel.

4   Place the sample application paper pieces on a clean glass plate and pipette protein sample on to each piece of paper avoiding cross-contamination. Place the sample applicators on the gel surface 1 cm from the acidic electrode strip.

5   Position the lid on the cell and ensure that the electrodes make a good contact with the strips.

6   Carry out the electrophoresis at the highest voltage compatible with the electrophoresis cell. Consult the manufacturer's recommendation on power settings.

4.3.2 Staining for the protein bands

Most of the staining methods employed for SDS–PAGE gels can be modified for application to IEF gels. The main requirement is to fix the IEF gel.

Protocol 13

Staining IEF gel for protein bands

Reagents

- Methanol, trifluoroacetic acid, sulfosalicylic acid, and water (30:10:3.5:56.5, by vol.)

- Methanol, trifluoroacetic acid, and water (30:12:58, by vol.)

Method

1   Fix the IEF gel in methanol, trifluoroacetic acid, sulfosalicylic acid, water for 1 h.

2   Wash the gel with several changes of methanol, trifluoroacetic acid, water for a minimum of 2 h.

3   Follow this by staining using the methods described before (*Protocols 9* and *10*).

4.4 Glycoprotein analysis

Various post-translation modifications such as glycosylation, carboxyl methylation of glutamate/aspartate residues, and phosphorylation can lead to electrophoretic and chromatographic heterogeneity of a pure proteins. For many proteins isolated from eukaryotes, glycosylation heterogeneity is probably the major contributor to the observed heterogeneity of the molecule. Although a wide variety of methods have been applied for the detection of glycoproteins, specific staining for the glycoprotein following electrophoresis is the simplest and most reliable. The method shown in *Protocol 14* is based on periodate oxidation of oligosaccharide attached to the protein followed by staining with Schiff's reagent (14, 15).

Protocol 14

Staining for glycoproteins

Reagents

- 5% phosphotungstic acid
- Methanol, acetic acid, and water (7:14:79, by vol.)
- 1% periodic acid
- 0.5% sodium metabisulfite
- Schiff's base

Method

1 Carry out SDS–PAGE as described in *Protocol 9*.

2 Soak the gel with gentle rocking in 5% phosphotungstic acid in 2 M HCl for 90 min at ambient temperature.

3 Wash the gel with methanol, acetic acid, water for 1 h. Repeat with fresh solution.

4 Soak the gel in 1% periodic acid and 14% trichloroacetic acid for 1 h.

5 Excess periodic acid is removed by washing the gel with 0.5% sodium metabisulfite in 0.1 M HCl.

6 Immerse the gel in Schiff's reagent on an ice bath in dark until pink bands develop gradually.

Note: a more sensitive method for glycoprotein detection uses labelled lectins in a manner similar to Western blotting (15).

4.5 Western blotting

Gel electrophoresis provides information on molecular weight and p*I* of the individual protein components present in given sample. By transferring the proteins separated by gel electrophoresis onto an immobilizing membrane allows these proteins to be probed for further characterization with specific reagents (e.g. antibodies). This approach can be adapted for determination of antigenic purity with antibodies directed against the product or the contaminants (provided contaminant specific antibodies are available). A number of methods, such as *Protocol 15*, have been described (16) but optimal conditions have to be determined empirically and will vary from sample to sample.

Several transfer membrane types (nitrocellulose and polyvinylidene difluoride) are commercially available. Equipment for transferring the gel electrophoresis separated proteins to the membranes is also available in a variety of formats. For quantitative procedures, the tank transfer systems are preferred but for large number of samples and qualitative results the semi-dry transfer electroblotters are suitable.

Protocol 15

Electroblotting and immunodetection

Equipment and reagents

- See *Protocol 8*
- Transfer buffer: dissolve 30 g Tris and 140 g glycine in 1 litre of water
- Tris-buffered saline (TBS): 10 mM Tris–HCl pH 7, 0.9% NaCl

Method

1 Carry out SDS-PAGE separation as described in *Protocol 8*. Include pre-stained markers (e.g. Amersham's rainbow markers) which can aid the monitoring of the electrophoretic separation and of the blotting procedure.

2 Soak the transfer membrane, four sheets of blotting paper, and the foam pads in the transfer buffer. Give hydrophobic PVDF membranes a short rinse in 100% methanol, followed by water and then buffer to aid wetting.

3 Assemble the transfer unit by first placing the foam pad followed by two pieces of the blotting paper, gel, membrane, two sheets of blotting paper, and the foam pad. Insert the assembled sandwich into the transfer unit with gel facing the cathode and the membrane on the anode side. Fill the chamber with chilled transfer buffer.

4 Close the lid and connect the electrodes to the power supply. Carry out the blotting at 1.2 A for 2–3 h.

5 On completion of the transfer, place the membrane in a tray with side which was in contact with the gel facing up. Incubate the membrane with blocking solution of 3% dried skimmed milk dissolved in TBS at room temperature for 5–18 h. Use a rotary shaking platform for best results.

6 Wash the membrane with TBS twice for 30 min.

7 Dissolve the primary antibody in the blocking solution, add sufficient volume to just cover the membrane, and incubate at room temperature for 5–6 h. The amount of antibody required will depend on the antibody titre and the nature/purity of the antibody.

8 Wash the membrane with TBS as before, then further wash for 5–6 h.

9 Detect the bound antibody by incubating with a secondary labelled antibody (see ELISA protocol). Alternatively, the primary antibody is probed using ^{125}I-labelled protein A or G followed by detection using a photographic film.

4.6 Chromatographic methods

4.6.1 Gel filtration

Gel filtration chromatographic analysis is based on the fractionation of proteins based on the relative size of the molecules and is one of the simplest methods for detection of impurities that differ in size from the target protein molecule. Since the method is non-destructive, intact molecules are obtained, the amount of material required will depend on the sensitivity of the assays used to detect the impurities. Full detailed instructions for the use of gel filtration matrices and columns are supplied by most suppliers of this technology (for detailed description see ref. 17). It is useful to calibrate the column with known molecular weight markers, which will allow a test for impurities as well as their molecular weights. In practice, during protein purification gel filtration is often used as a final polishing step and thus, can be used not only to remove the impurities but also to confirm the relative purity of the target molecule.

In gel filtration the sample is applied at the inlet of the column and is eluted in an isocratic mode with any mobile phase which is compatible with the target biomolecule and minimizes the interaction of the molecules with the column packing. The main interactions of the dissolved molecules with the packing are hydrophobic and electrostatic in nature. The ionic strength of the mobile phase should be high enough (> 0.1 M salt) to counter the effects of electrostatic interactions but not too high (< 0.5 M salt) to promote hydrophobic interactions.

As the applied band moves down the column, the large excluded molecules move down with the mobile phase in a volume equal to the volume between the packed particles (void volume). Smaller molecules will enter the pores in the particles and elute in volume equal to the volume occupied by the pores and the volume between the packed beads. The intermediate size molecules elute in between and the range of molecular sizes separated is dependent on the pore size distribution of the packed beads. Therefore, selection of the appropriate grade of the separation media is critical for achieving high resolution.

4.6.2 Ion exchange chromatography

High performance ion exchangers (e.g. the Mono Q and S from Pharmacia and the Poros range available from PerSeptives Biosystems) are often used to ascertain the purity of the final product. The high resolution and the short separation times achieved with these supports make them ideal for such analysis. Purity analysis by IEC exploits the fact that proteins are ampholytes as they contain both positive and negative charges and the net charge is dependent on the pH of its environment. Therefore, by carrying IEC at several different pH values (pH Map), the impurities with a small pI difference can be resolved from the target molecule. For purity analysis purposes, a single small particle HPLC column, with a shallow, well-optimized gradient is used.

The selection of an ion exchange column is usually based on the pI and pH stability of the protein molecule of interest. At pH values above the pI, the molecule will have a net negative charge and anion exchange column should be used. Below the p*I* the molecule is positively charged and a cation exchange column should be used. It should be noted however, that p*I* value reflects the overall charge distribution on the protein molecule, but the tertiary structure of the protein molecule does not allow all the charged residues to interact with the ion exchangers simultaneously. The interaction is often localized at a site with an abundance of similar charge. Proteins can, therefore, be retained by both types of ion exchangers at or beyond their isoelectric points. Therefore, isoelectric points are only used as a guideline in determining the type of ion exchanger to be used.

Elution is often carried out at a constant pH with increasing ionic strength to displace the bound protein and sodium chloride is the commonly used salt. The purity of a protein preparation is indicated by the symmetrical elution peak and the lack of other minor peaks in the elution profile.

4.6.3 Reversed-phase chromatography

RPC is the most common method for analyses of most types of molecules. The separation is based on differences in hydrophobicity of the molecules and this is the method of choice for the analyses of purity for proteins. However, the strongly hydrophobic bonded phase and the use of organic solvent elution conditions often lead to denaturation of proteins.

Many of the commonly used reversed-phase matrices are silica based with bonded hydrocarbon chains or polymeric support matrix, such as polystyrene divinylbenzene.

The hydrocarbon chain most effective for proteins is in the range C_4 to C_8. For analytical purposes, non-porous supports provide a rapid method for ascertaining the purity of a protein preparation.

The selectivity in RPC is achieved by the selection of an appropriate mobile phase. Ion-pairing agents are used to enhance the interaction of charged groups on the sample molecule with the hydrophobic bonded phase. Trifluoroacetic acid is the most commonly used ion-pairing agent because of its ability to dissolve proteins and lack of absorbance at wavelengths used to monitor for proteins. Selectivity and sensitivity are often improved by using other ion-pairing agents, such as hydrochloric acid, formic, phosphoric, and acetic acids. The selection of the organic component is based on the protein solubility and the solvent viscosity. The solvents used are water-miscible, such as acetonitrile, methanol, and 2-propanol.

After sample application, the elution is performed by reducing the polarity of the mobile phase which causes the bound molecules to elute off the packed bed. The gradient elution protocol cannot be predicted and has to be experimentally determined. As a starting point, a linear gradient from 5% organic to 95% organic over 30 min is used to examine the resolution of the target molecule from the impurities. The gradient is then optimized for higher resolution.

5 Other biochemical and biophysical methods

5.1 Protein mass spectrometry

Recently, protein analysis by mass spectrometry has provided a massive impetus in the field of complex protein analysis and is far more accurate than the techniques described earlier in this chapter. Recent techniques permit not only the determination of molecular weight of intact proteins but also primary structure analysis, rapid protein identification by its peptide mass fingerprint, and amino acid sequencing by mass spectrometry. Developments in the technique have changed the method such that it is increasingly used by skilled biochemists for protein characterization. These developments have recently been presented in detail by the experts in the field (18).

References

1. Mohan, S. B. (1992). In *Methods in molecular biology. Vol. 11. Practical protein chromatography* (ed. A. Kenney and S. Fowell), p. 301. The Humana Press, Totowa, NJ.
2. Harris, E. L. V. and Angal, S. (ed.) (1989). *Protein purification applications: a practical approach*. IRL Press, Oxford.
3. Kirshenbaum, D. M. (1976). In *Handbook of biochemistry and molecular biology* (ed. G. D. Fasman) (3rd edn), Vol. 2, p. 383. CRC Press, Cleveland, Ohio.
4. Stoscheck, C. M. (1990). In *Methods in enzymology* (ed. M. P. Deutcher),Vol. 182, p. 50. Academic Press Inc.
5. Rhodes, D. G. and Laue, T. M. (1990). In *Methods in enzymology* (ed. M. P. Deutcher), Vol. 182, p. 555. Academic Press Inc.
6. Pace, N. C. and Schmid, F. X. (1997). In *Protein structure: a practical approach* (ed. T. E. Creighton), p. 253. IRL Press, Oxford.
7. Smith, P. K., Krohn, R. I., Hermanson, G. T., Mallia, A. K., Gartner, F. H., Provenzano, M. D., *et al.* (1985). *Anal. Biochem.*, **150**, 76.
8. Stoscheck, C. M. (1987). *Anal. Biochem.*, **160**, 301.
9. Pepper, D. S. (1992). In *Methods in molecular biology. Vol. 11. Practical protein chromatography* (ed. A. Kenney and S. Fowell), p. 135. The Humana Press, Totowa, NJ.
10. Scheidtmann, K. S., Reinartz, S., and Schlebusch, H. (1997). In *Protein structure: a practical approach* (ed. T. E. Creighton). IRL Press, Oxford.
11. Parker, C. W. (1990). In *Methods in enzymology* (ed. M. P. Deutcher),Vol. 182, p. 721. Academic Press Inc.
12. Chrambach, A. (1985). *The practice of quantitative gel electrophoresis*. VCH, Weinheim.
13. Andrews, A. T. (1986). *Electrophoresis: theory, techniques, and biochemical and clinical applications*. Oxford University Press.
14. Garfin, D. E. (1990). In *Methods in enzymology* (ed. M. P. Deutcher),Vol. 182, p. 459. Academic Press Inc.
15. Gerard, C. (1990). In *Methods in enzymology* (ed. M. P. Deutcher),Vol. 182, p. 529. Academic Press Inc.
16. Leach, B. S., Collawn, J. F., and Fish, W. W. (1980). *Biochemistry*, **19**, 5734.
17. Stellwagen, E. (1990). In *Methods in enzymology* (ed. M. P. Deutcher),Vol. 182, p. 317. Academic Press Inc.
18. Jensen, O. N., Shevchenko, A., and Maan, M. (1997). In *Protein structure: a practical approach* (ed. T. E. Creighton), p. 29. IRL Press, Oxford.

Chapter 4
Clarification techniques

Ian Reed
AEA Technology plc, Didcot, Oxfordshire OX11 0RA, UK.

Duncan Mackay
AEA Technology plc, Didcot, Oxfordshire OX11 0RA, UK.

1 Introduction

Proteins can be produced by a number of different routes such as fermentation, tissue culture, and by extraction from plasma or plants. Whatever route is chosen, the raw protein-bearing stream is likely to be a complex mixture containing both dissolved species and particulate material. The target protein will be present at very low concentration and with a host of contaminants such as cells or cell debris, DNA, proteins and polysaccharides, and a large quantity of water. Such a mixture is very difficult to treat using the highly selective processes that are required to obtain the target product at high purity since the presence of particulate material impairs their function. The first challenge of protein purification is therefore to convert the complex fermentation broth which is a mixture of dissolved and suspended solids into a form that is amenable to further purification. Although there is much interest in direct recovery of protein from such materials, the most frequent first step currently is to clarify the raw protein source to remove suspended matter. It is then possible to use a range of highly selective techniques to purify the target protein.

There are a number of clarification techniques that can be adopted and the choice of which to use depends on both the source of raw feed and the scale of operation. There are two main classes of process; sedimentation and filtration. Sedimentation can be carried out under normal gravity conditions or, as is almost always the case for biological streams, using a centrifuge. Filtration can be performed using either conventional filter media or using membrane filters for removal of finer particles. The aim of this chapter is to describe these methods, and their underlying principles, the advantages of each are discussed, and examples of equipment are presented. Practical advice is presented on how and when to use each technique.

2 Sedimentation

Sedimentation processes operate primarily on the basis of density differences between the various components of a mixture. They are most commonly applied to suspensions of solid in liquid, but also to disengage immiscible liquids. If there is no density difference between particulates and the suspending medium, sedimentation cannot occur. However, where there is a density difference, the rate of sedimentation will be influenced by a number of factors including particle size, shape, and concentration as well as the magnitude of the density difference.

Two measures can be taken to increase the settling rate. By far the most effective is to use centrifugal forces to increase the driving force, which can be achieved using a centrifuge. At the high centrifugal rates generated by some laboratory centrifuges, only tiny density differences are required to effect a separation. Further improvements can be obtained by flocculating or coagulating the cells, but neither method will be effective if there is no density difference in the first place.

2.1 Principles of sedimentation

Gravity sedimentation is rarely used for the clarification of cell suspensions because of the low settling velocities and the consequent large size of equipment used. However, the principles are important since they also apply to centrifugal sedimentation which is probably the most commonly used method.

The key parameter in determining the rate of separation is the settling velocity and a particle settling within a medium is subject to a number of forces. Gravitational forces are counteracted by buoyancy and drag forces and the particle reaches maximum or terminal velocity when these forces are equal. There are many theories of settling and the following equation gives a commonly used result:

$$u_p = \frac{u_c}{e} = \frac{d^2(\rho_s - \rho_c)g}{18\,\mu_c} \qquad [1]$$

where u_p is settling velocity of a particle relative to the fluid (m/s), u_c is settling velocity of the suspension (m/s), e is volume fraction of fluid in the suspension (−), d is particle diameter (m), r_s is density of solid phase (kg/m^3), r_c is density of suspension (kg/m^3), g is gravitational acceleration (m/s^2), and m_c is viscosity of suspension (Ns/m^2).

Equation 1 indicates that settling rate is a function of the gravitational force, density difference, particle diameter, solution viscosity, and solution concentration. For most practical purposes settling rate can only be increased by increasing the particle diameter or the g force. The former can be achieved by coagulation and flocculation of the particles and the latter by using a centrifugal separator. For certain specialized separations, it is sometimes possible to increase density difference by binding particles to a dense substrate such as silica.

2.1.1 Coagulation

The terms coagulation and flocculation are often used interchangeably, but are in fact quite different phenomena. Coagulation is the gradual increase in particle size resulting from like particles joining to form larger particles. Flocculation is the bridging of particles by polymers or large particles of another material.

Particles suspended in a medium nearly always carry an electrical charge. This can be caused by the adsorption of ions from the solution or from ionization of the particles themselves. These charges cause the particles to repel one another and prevent coagulation. At shorter distances van der Waals forces cause interparticle attraction leading to a potential energy curve similar to that shown in *Figure 1*. Normally particles lie outside the primary maximum and thus repel. Occasionally two particles will meet with sufficient kinetic energy to overcome the repulsion and thus coagulate. The higher the charge on the particles the less likely it is that coagulation will occur. Coagulation can therefore be promoted by reducing the charge (and the potential barrier) ideally to zero. This can be achieved in a number of ways.

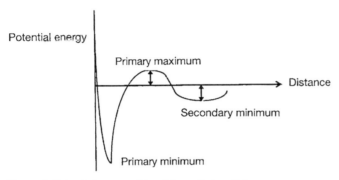

Figure 1 Potential energy of two interacting particles.

One common technique is to add multivalent ions with a charge opposite to that on the particles. Since most particles carry a negative charge in suspension, aluminium ions (Al^{3+}) are particularly effective, although they are not generally appropriate for protein purification because of their toxicity. Simply increasing the ionic strength of the suspending medium can also aid coagulation by shielding the charges on particles, but the method can hamper subsequent purification steps.

Probably the most widely used means of coagulating biological systems is by pH adjustment. Many biological polymers are amphoteric, that is they contain both acidic and basic portions. As a result their surface charge is a function of pH. At low pH they are negatively charged and positively charged at high pH. There is thus an intermediate pH, known as the isoelectric point, where the net charge is zero. At this point there is no repulsion of particles and conditions are ideal for coagulation. Biomolecules have a characteristic isoelectric point and can be precipitated selectively by adjusting to the correct pH.

2.1.2 Flocculation

In flocculation, average particle diameter is increased through the action of flocculating agents that act as bridges between particles. Flocculating agents are generally polyelectrolytes and function by a two-step process. The first step is the neutralization of surface charges on the suspended particles and is followed in a second step by linking of particles to form larger aggregates.

Flocculating agents are effective at much lower dose rates than coagulants and are the preferred method of aggregation for biological systems. It is important not to overdose as this can lead to reversal of the charges on particles thus hindering particle aggregation.

A wide range of flocculating agents is available commercially and testing is required to determine the most effective flocculant and dosage for each application.

2.2 Gravity sedimentation

Gravity sedimentation is very cheap for large scale applications; industrial devices can be in excess of 10 m across. However, separation is generally rather slow, leads to relatively dilute solids streams, and the clarified liquid still carries some fine particulates. This is particularly apparent in biotechnology applications where the density difference between the particulates, usually cells or cell debris, and the suspending medium, water, is nearly always very low. Unaided gravity sedimentation is rarely used at a laboratory scale. Even where biomass settles rapidly, centrifugal separation is often used since it reduces settling times, increases separation efficiency, and reduces the volume of sediment. The latter factor is particularly important due to the increased yield of product in the supernatant. Increasing separation efficiency makes subsequent purification steps more efficient and reliable.

3 Centrifugal sedimentation

Enhanced gravity settling is a widely used method for the separation of biomass (cells, cell debris, etc.) from fermentation liquor or buffer. Disc stack, tubular, and scroll decanter centrifuges are common equipment in the biotechnology industry. The recovery efficiency depends on the following; the settling rate of the particulates, the residence time within the centrifugal field, and the settling distance.

As in gravity sedimentation, the settling rate can be increased by increasing the biomass size, increasing the density difference between the biomass and the suspending liquid, or by decreasing the viscosity of the liquid. Separation can also be improved by increasing the g force. The residence time can be increased, and hence the recovery efficiency, by reducing the flow rate through the centrifuge. The settling distance can be reduced and recovery efficiency increased, for example by inserting discs into the separation chamber.

The scale-up from laboratory to pilot to full-scale operation of centrifuges relies on the Sigma concept. This relates the biomass recovery to a factor Q/Σ, where Q is the volumetric flow rate of the process stream and Σ is a geometric parameter characteristic of a particular centrifuge.

3.1 Centrifuge equipment

3.1.1 Tubular

Tubular centrifuges are basically vertically mounted long cylinders with an inlet and outlet for the liquid. The solids deposit on the inner walls and are discharged at intervals. Tubular centrifuges operate at high g forces (62 000) and are used for clarification. They are limited to process streams with relatively low biomass content.

3.1.2 Disc stack

The design of these centrifuges incorporates numerous discs (typically between 30 and 200) which divide the sedimenting bowl into separate settling zones. Disc stack centrifuges operate in either batch or continuous mode at g forces typically in the range 4000–10 000 and can be used with process streams with biomass content up to 30% (v/v).

The various types of disc stack centrifuge can be classified according to the method of solids discharge and include solid bowl, opening bowl (see *Figure 2*), nozzle valve, and slot.

3.1.3 Scroll decanter

Scroll decanter centrifuges (*Figure 3*) consist of a solid bowl, tapered at one end, and a close fitting helical screw which rotates at a slight differential speed to the bowl. The screw continuously discharges sedimented solids from the bowl. The

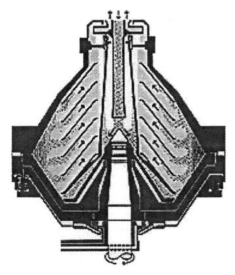

Figure 2 Opening bowl disc stack centrifuge.

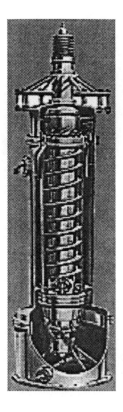

Figure 3 Vertical scroll decanter centrifuge.

settling distance can be altered by the height of the pool of clarified liquid in the bowl. The g forces are in the range 4000–10 000 and streams with a biomass content up to 80% (v/v) can be processed.

3.2 Laboratory centrifuge test work

To determine the best conditions for clarification, the sedimentation rate of cells or other biomass should be measured under gravity or enhanced gravity conditions. In practice, the former may be of little value due to very low settling velocities. Nevertheless, suspensions that settle under gravity at least 70% of the settling distance within 30 min and form a distinct settling interface will be simple to separate in a centrifuge.

The likely performance of centrifugal sedimentation can also be estimated by calculating a particle size, x_{max}, representing the smallest particle which can be 100% recovered.

$$(x_{max})^2 = [18 \cdot \mu/(g \cdot \Delta\rho)] \cdot [(1 - \Phi)^{4.6} / E] \cdot [Q/\Sigma] \qquad [2]$$

where μ is the liquid viscosity (Ns/m^2), g is the acceleration due to gravity (m/s^2), $\Delta\rho$ is the density difference between the solid and liquid phases (kg/m^3), Φ is the solids volume fraction (−), E is the efficiency of the centrifuge (variable but typically 40–55% for disc stack machines), Q is the volumetric flow rate of the feed (m^3/s), and Σ is the characteristic Sigma value of the centrifuge.

Given a feed suspension flow rate and the centrifuge parameters, a value for x_{max} can be determined. Comparison with the particle size distribution of the feed will provide an initial assessment of likely performance. If the calculated x_{max} lies too far to the right of the particle size distribution of the feed, then flocculation should be considered.

3.2.1 Test apparatus

Simple gravity jar tests could be carried out but it is likely that the low settling rates will make this a time-consuming process. There are commercially available jar test apparatus, but it is often sufficient to use test-tubes or measuring cylinders with simple inversion/shaking.

Tests are best performed using a laboratory swing-out rotor centrifuge. An angle rotor centrifuge could be used, but would be less satisfactory because the settled biomass would form a sloped sediment making settling distance difficult to calculate and the g force would have to be averaged.

If a small disc stack centrifuge is available (e.g. Alfa Laval Gyrotester) and there is sufficient process stream then additional useful performance data can be obtained.

3.2.2 Test procedures

The purpose of the tests is to measure the clarity of the supernatant, the sediment volume, and the nature of the sedimented solids (moisture content and flow properties) as a function of centrifugation time and g force.

i. Jar test

This gravity settling test described in *Protocol 1* is the classical method used in the water treatment industry for sizing gravity thickeners and clarifiers. The test is mainly for providing information on the most suitable flocculation conditions but it can also give an estimation of the settling rate of an untreated suspension.

Protocol 1

Assessment of suspension settling properties using a jar test

Equipment and reagents

- Jar
- Stoppered test-tube or measuring cylinder
- Stopwatch
- Ruler
- Sample of suspension

Method

1 Place a representative sample of the cell suspension (around 50–100 ml) into the test vessel.

2 Seal top of container and invert repeatedly until the suspension is homogeneous.

Protocol 1 continued

3   Place the test vessel on a level surface and start the stopwatch.

4   Measure and record at regular intervals the height of the interface between the suspensions and the clear liquid.

The key parameter from this test is the settling velocity as indicated by the movement of the suspension/supernatant interface.

ii. Centrifuge test

In practice, the settling rate for gravity sedimentation is likely to be very low. More useful separation data can be obtained using a simple laboratory centrifuge as described in *Protocol 2*. Round-bottomed 50 ml centrifuge tubes should be used. A known weight or volume of suspension is added to each tube, the tubes spun at a specified g force, and a tube removed at different times for analysis.

Protocol 2

Assessment of suspension settling properties using a centrifuge

Equipment and reagents

- Vacuum oven
- Swinging rotor centrifuge with controllable rotation speed
- 50 ml graduated round-bottom centrifuge tubes
- Sample of suspension

Method

1   Take a sample of the suspension and measure the solids content by evaporating to dryness in a vacuum oven at 60 °C.

2   Weigh each centrifuge tube and place a measured weight or volume of homogeneous suspension into each tube (the number depends on the design of the centrifuge).

3   Load the centrifuge and run at a measured rotation speed and duration.

4   Remove the centrifuge tubes and decant the supernatant liquid. Measure the volume and dry solids content as in step 1.

5   Measure the weight of each tube plus sediment to determine mass of settled solids. Take a sample of the sediment and measure solids/moisture content as in step 1.

6   The test yields the following parameters: weight or volume of suspension used (tube tests), biomass content of suspension, centrifuge rotor rotational speed (from which g force can be calculated), biomass content of supernatant, sediment volume, and sediment moisture content.

If a small scale disc stack centrifuge is used, a similar test can be performed by operating the centrifuge as per the manual and taking samples of the feed, supernatant/centrate, and discharged solids at various times for analysis.

For disc stack tests calculate the parameter Q/Σ which represents the settling velocity of the smallest particle that can be totally recovered. Q is the volumetric flow rate (m³ s⁻¹) and Σ is the Sigma factor for a disc stack centrifuge which can be calculated from the following equation:

$$\Sigma = {}^{2}/_{3}\left(\frac{\omega^2}{g}\right)\pi n\,(r_2{}^{3} - r_1{}^{3})\cot\theta \qquad [3]$$

where ω is the rotational speed (rad s⁻¹), g is gravitational acceleration (m s⁻²), n is the number of discs, r_2 is the distance from the rotational axis to the outer edge of the discs (m), r_1 is the distance from the rotational axis to the inner edge of the discs (m), and θ is acute angle the discs make to the rotational axis (rad).

Where possible, centrifuge tests should be performed at a range of conditions to determine optimum performance. The main parameters that can be varied in the centrifuge tests are: spinning (residence) time, g force, suspension flow rate (for disc stack centrifuges), settling distance, and sediment discharge rate (for disc stack centrifuges).

3.2.3 Test data interpretation

Correct analysis of the test data will provide a measure of the sedimentation properties of the suspension solids, i.e. biomass. A quantitative assessment of the solids recovery performance can be made by plotting percentage solids recovery against g force (for tube tests) or Q/Σ (for disc stack tests):

$$\text{Solids recovery} = 100\left[1 - \frac{C_S}{C_F}\right] \qquad [4]$$

where C_S is the solids content of the supernatant (kg/m³) and C_F is the solids content of the suspension (kg/m³).

The relationships between biomass recovery and the operating conditions allow the separation efficiency to be assessed.

By plotting sediment moisture content and/or sediment volume (for tube tests) against g force (for tube tests) or Q/Σ (for disc stack tests) the extent of biomass thickening can be determined.

3.3 Suspension conditioning

The test data may indicate that centrifugal recovery will be inefficient because of low settling rates. In such cases it may be possible to improve performance by conditioning of the suspension. For suspensions where the gravity settling rate is less than about 10^{-5} m min⁻¹ or the product of g force and spinning time (for centrifuge tube tests) is greater than 2×10^7 sec settling will be difficult and flocculation of the suspension is recommended.

Conditioning of the suspension can be used to increase the biomass sedimentation rate, the biomass recovery efficiency, and the de-watering of the biomass. Conditioning can be achieved with or without additives. Methods not requiring additives include:

(a) Alteration of the temperature—flocculation can occur because of changes in the interactions of surface macromolecules.

(b) Alteration of the pH—changes in the degree of ionization of cell surface groups can affect the electrostatic interaction between surfaces and also the conformation of macromolecules which extend into the bulk solution.

(c) Changes in the fermentation—the biomass properties and the composition of the fermentation medium will vary depending on the metabolic condition of the cells.

(d) Autoflocculation—cells may flocculate naturally by virtue of the properties of their surface chemistry.

Additives can also be used to increase mean particle sizes:

(a) Inorganic salts—ions can bring about aggregation (coagulation) of cells as the result of changes in the electrostatic interaction between surfaces.

(b) Polymers—uncharged polymers and polyelectrolytes are able to aggregate (flocculate) cells because surface adsorption can give rise to surface charge neutralization or surface bridging.

In general, polymeric flocculants represent the preferred choice for conditioning. The sedimentation tests described in *Protocols 1* and *2* should be repeated on the conditioned broth to determine changes in settling rates.

4 Filtration

4.1 Background

Filtration is ubiquitous in clarification of all types of liquid. The basic principles are very simple although the practicalities involved in operating large scale devices are often less so. The liquid to be clarified is passed through a porous barrier. Large particles are retained at the filter surface or within the depth of the filter medium, solvent and some small particles pass through the filter. The process is pressure-driven and unlike sedimentation works well even when there is no density difference between the particulates and the suspending medium.

Filtration is used in a wide variety of solid/liquid separation applications and this is reflected in the large range of filtration equipment that is available. The design of filtration equipment is influenced by the physical properties and concentration of the particles, the type of solvent, scale of operation, the desired product, and the mode of filtration. There are three principle modes of filter operation; dead-end (sometimes called flow-through), cross-flow, and depth or bed filtration.

4.1.1 Depth filtration

Depth filters consist of a packed bed of large relatively dense particles often with a graded particle diameter through the filter. The liquid to be filtered is trickled through the bed by gravity and because of their greater momentum, large particles in the feed stream impact upon the packing and are held within the bed; smaller particles pass through. The particles trapped by the bed can be recovered by periodically back-flushing. Deep bed filtration is cheap but not very efficient and is almost always employed at very large scale for the clarification of dilute suspensions such as in water treatment. It is not a practical technique for laboratory scale applications.

4.1.2 Dead-end filtration

Dead-end filtration is the simplest form of filtration (see *Figure 4*). The liquid is presented to a static filter medium under pressure or in some cases drawn through the filter by an applied vacuum. Usually the filter medium contains pores that are larger than the particles to be filtered. Initially solid/liquid separation can occur either at the filter surface or within the filter medium and deposition of particles causes pores to narrow until all (large) particles are retained at the filter surface. A filter cake then builds-up as retained material accumulates and in many cases this filter cake is essential to the separating properties of the filter. At laboratory scale almost all filtration, except for membrane filtration, is carried out in dead-end mode and the majority of this section therefore concentrates on this form of filtration.

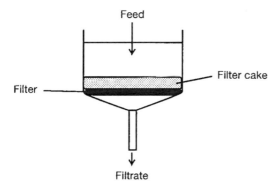

Figure 4 Dead-end filtration (Buchner funnel).

4.1.3 Cross-flow filtration

Conventional filtration media similar to those used in dead-end filtration can be used in cross-flow mode. In cross-flow filtration the feed is pressurized and fed tangentially to the filter surface at velocities of up to 6 m/s (see *Figure 5*). Unlike dead-end filtration the pore diameter of the filter medium is usually less than that of the particles to be filtered and separation occurs at the filter surface. The main advantage compared to dead-end filtration is that the shearing action of the transverse flow reduces the thickness of the filter cake and thus increases, in principle, the filtration rate. However, where the process stream contains deformable particles, such as cell suspensions, this is not always realized as the cake that forms in cross-flow is usually much less permeable than that formed

61

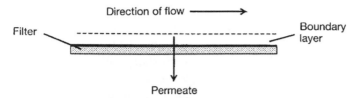

Figure 5 Schematic of cross-flow filtration.

in dead-end filtration. The obvious disadvantage of cross-flow filtration is that it requires more sophisticated plant and has a high energy consumption. At small scale it is rare to use conventional filters in cross-flow. The principle is mostly associated with membrane filters, which are discussed in detail in Section 5.

4.2 Principles of dead-end filtration

The simplest way to operate filters is at a constant pressure. As a cycle progresses a cake layer builds-up and the filtrate flow declines. This section looks at filtrate volume against time for the case of a suspension of incompressible particles filtered at constant pressure.

Assuming that the filter medium itself offers minimal resistance to solvent flow, the filtration rate per unit of filter area (left-hand side of the equation) can be represented by Equation 5:

$$\frac{1}{A}\frac{dV}{dt} = \frac{-\Delta P}{r \cdot \mu \cdot \ell} \tag{5}$$

where V is filtrate volume (m^3), t is time (s), A is filter area (m^2), ΔP is pressure differential across filter (N/m^2), r is specific resistance of filter cake (m^{-2}), μ is fluid viscosity (Ns/m^2), and ℓ is cake thickness (m).

In dead-end filtration of an incompressible material, the volume of the filter cake is directly proportional to the filtrate volume and this can be represented by:

$$V_{cake} = \ell \cdot A \equiv v\,V \tag{6}$$

where V_{cake} is volume of filter cake (m^3), and v is volume of filter cake per unit volume of filtrate (–). Substituting for l in Equation 6 gives:

$$\frac{1}{A}\frac{dV}{dt} = \frac{-\Delta P}{r \cdot \mu\,v \cdot V}\,A \tag{7}$$

rearranging and integrating:

$$\int_0^V V dV = -\int_0^t \frac{\Delta P \cdot A^2}{r \cdot \mu \cdot v \cdot V}\,dt \tag{8}$$

$$\Rightarrow \qquad \frac{V^2}{2} = \frac{r \cdot \mu \cdot v}{A^2(-\Delta P)}t \tag{9}$$

Thus, the cumulative filtrate volume varies as $t^{\frac{1}{2}}$ and the filtrate flow rate as $t^{-\frac{1}{2}}$.

4.2.1 Filter aid

Two significant problems can occur with filtration; low filtrate flow rate and poor clarification. These difficulties often arise where the suspended solids are compressible and thus form low porosity, high resistance filter cakes. Deformability of particles can also lead to reduced retention by the filter medium. Typical fermenter broths can suffer from both problems, but performance can often be improved by the addition of a filter aid.

Filter aids are inert materials that operate solely by mechanical means and there are two modes of action. First, filter aids may be used as a pre-coat to create a porous barrier between the filter cloth and the cake. This reduces the effective pore size of the filter medium and prevents pore blocking by the suspended particles. The second mode of action is as a body feed where the filter aid is mixed with the suspension. It improves filtration rates by promoting the formation of an open, highly permeable filter cake.

For best effect filter aids should be inert, incompressible, insoluble, and irregularly shaped. Diatomite, which comes in a number of forms such as diatomaceous earth, diatomaceous silica, and kieselguhr, is by far the most industrially important. It is composed principally of silica and is used in a wide range of industries ranging from antibiotics production to lube oils and whisky. Other filter aids include asbestos, cellulose, and perlite.

There are two main drawbacks to using filter aids. The first is the additional cost, which may not be great for materials such as diatomaceous earth, but the second, disposal, can present greater problems particularly at large scale. The mixture of cells and filter aid can be difficult and expensive to dispose of and in some cases, operators prefer to use membrane processes in order to avoid this problem.

4.2.2 Media for dead-end filtration

Filters can be made from a wide range of materials using a number of different production methods. The requirements for a good filter medium are good mechanical strength, high porosity, consistent pore dimensions, low resistance to flow, good chemical stability, and low cost. At laboratory scale, the most familiar medium is probably filter paper. Paper is cheap and effective for small volumes of liquid, but poor mechanical properties and relatively low permeability preclude its use for large scale applications. At large scale, woven fabrics are the most commonly used filter materials due to their low cost and fair mechanical properties. Because of their flexibility fabrics are widely used in rotary filtration devices such as rotary drum vacuum filters (RDVF) or belt filters. Capsule filters used at small to medium scale generally use plastic filter media. These are most commonly non-woven materials in which a mat of randomly arranged fibres is bound together either by partial melting or gluing. For high temperature applications or where the liquids are particularly corrosive it is possible to make filters from sintered metals or metal oxides, but these materials are expensive.

Unlike most membrane filters (described in Section 5) laboratory scale filters are cheap enough to be single use. It is, therefore, rare to use chemical cleaning on conventional filter media.

4.3 Laboratory scale filtration equipment

At industrial scale, the range of filtration equipment is extensive. Although the basic principles are simple, the number of variations in design used to accomplish the task is considerable. In most cases these are intended as a means of turning what is essentially a manually intensive batch operation into an automatic or continuous process. At laboratory scale, the requirements are much simpler in that process volumes are much smaller and batch operation is preferred to continuous. As a result, the choice of systems is correspondingly much more limited.

Small scale filters are almost always surface or cake filters. Although in some cases there may be an element of depth filtration, this is never the primary mechanism of operation. Traditional depth type filters such as trickling bed filters are much too large for convenient small scale use. The following section looks at the type of filters available for laboratory operation and indicates where each might be used.

4.3.1 Buchner filters

Buchner funnels are probably the most familiar filters in a laboratory environment. A typical Buchner funnel has a flat porous base and vertical sides and in use is mounted in a conical flask with a side-arm for application of a vacuum. A suitable filter material, usually a filter paper, but muslin or wire meshes are also suitable, is fitted to the porous base so that all pores are covered. The liquid to be filtered is then poured into the funnel and collected in the flask. If necessary, it is possible to increase the filtration rate by applying a vacuum to the filtrate collection vessel.

Buchner funnels of appropriate size can be used to clarify volumes of several hundred litres. Quite high concentrations can be handled provided that solids form a high porosity cake. The method has been used to clarify fungal fermentation broths where the filamentous organisms produce an open, permeable cake. However, the biomass from many bacterial fermentations is slimy and may be more difficult to filter. The method has inherently poor containment and is not suitable for hazardous materials.

4.3.2 Bag filter

Bag filters are sock shaped devices made of fabric or porous plastic with a stiffened opening into which liquid is poured. Filtrate passes through the filter bag under gravity only. Bag filters can be a relatively rapid method of filtering volumes of around 10–20 litres and they tend to be quicker than Buchner funnels as a larger surface area is presented to the permeating liquid and the pore size is generally greater. However, the solids holding capacity is relatively limited and operation is inconvenient for large volumes.

4.3.3 Cartridge filter

Cartridge filters come in a range of sizes and can be equipped with either membrane or conventional filters. Filter housings are usually cylindrical vessels containing a flat sheet filter, pleated in order to increase the available surface area. In some designs, the filter element can be removed from the cartridge and replaced, but increasingly small scale devices are produced as an integral disposable unit. Cartridge filters have a high particle removal efficiency and because of the high surface area give high filtrate flow rates. However, the solids holding capacity is very limited and cartridges are best suited to filtering dilute streams. This filter design is widely used for polishing, i.e. removal of the last vestiges of particulates, and in-line sterilizing filtration. Cartridge filters are not suitable for clarification of fermentation broths.

4.3.4 Filter press

Filter presses are sophisticated devices widely used in industrial scale processes. Smallish units are available for pilot scale use. Typical designs consist of a set of plates mounted into a frame or press. Filter medium is stretched across the plates which are grooved to create channels for the filtrate to flow through. Once assembled the press has a manifold arrangement that allows the process stream to be fed under pressure between the plates. At the end of each operating cycle, the filtered solids are usually discharged by opening valves at either end of the filter plates. When this is not possible the press can be disassembled and solids recovered manually.

4.4 Application of filtration

The main advantages of conventional filtration are high filtration rates, relatively low cost, mechanical simplicity (not in all designs), relative ease of maintenance, and recovery from poor performance. The main drawbacks are low retention, poor containment, and the need to add filter aid in order to ensure good filtration. Often the clarification step is a preliminary to cell disruption and in such cases, it is desirable that the cell suspension remains sufficiently fluid to be passed through a homogenizer or bead mill. Dead-end filtration generally produces a filter cake that requires resuspension. This extra step can increase both the cost and time required for clarification, but is unlikely to be an issue at small scale. At laboratory scale, the most widely used types are Buchner funnels and capsule or cartridge filters.

5 Membrane processes

5.1 Background

For some time the term membrane process was applied to separations effected by the use of thin, selective, semi-permeable barriers. However, with the advent of ceramic and metal filters able to perform similar separation tasks this definition is no longer completely appropriate. Instead it is best to regard membrane filters as those that can operate in the submicron range and where separation occurs at the surface of the filter.

Using this definition, the main difference between membrane and conventional filtration is in the size of particles that can be retained. There is no precise distinction, but filters with pore sizes up to around 5 μm are generally regarded as membranes. Media with larger characteristic pore sizes are more usually thought of as conventional filters. The other significant difference is that in membrane filtration separation occurs at the surface of the filter whereas conventional filtration may occur by any of a number of mechanisms (including surface filtration).

The relatively small difference between membranes and conventional filters has some far reaching consequences for the ways in which these media are used.

(a) Membranes are more expensive to produce than conventional filters and, particularly at large scale, must be reused many times. Even for large laboratory applications (> 100 litres) membranes are frequently too costly to be considered as disposable items.

(b) Because of their small pore sizes membranes can have quite low treatment rates. Given the relatively high cost of the filters it is therefore particularly important to optimize operating conditions.

(c) Membranes operate at a molecular level. Interactions with macromolecules such as proteins can therefore affect the separating properties and the chemical properties of the filter material are therefore much more important than for conventional filters.

This section sets out the operating characteristics of membranes in sufficient detail to understand the main factors that influence behaviour and to identify suitable operating conditions. Protocols are provided to aid selection of membrane types.

5.1.1 Definition of terms

There are a number of special terms that apply to membrane processes and it is worthwhile to define these in order to aid understanding the processes.

(a) **Permeate**—the stream that permeates through the membrane (equivalent to filtrate in filtration).

(b) **Retentate**—the portion of the feed stream that is held back or retained by the membrane.

(c) **Flux**—a measure of the treatment rate, permeate flow rate per unit of membrane area usually expressed in the units litres per hour per square metre of membrane area (litre/m^2/h) or m/s.

(d) **Rejection**—the proportion of a particular species retained by the membrane. A rejection of 0.8 or 80% implies that 80% of the species is retained by the membrane and 20% appears in the permeate.

(e) **Molecular weight cut-off**—molecular weight cut-off is a measure of the separation properties of an ultrafiltration membrane. Ideally all molecules larger than the molecular weight cut-off should be retained by the membrane and all smaller species should permeate. In practice the distinction is not so sharp and the cut-off is usually defined as the molecular weight of spherical molecule for which the membrane gives a rejection of 90%.

(f) **Nominal pore size**—a measure of the separation characteristics of a microfiltration membrane; usually the diameter of the largest pore.

5.2 Principles of operation

Like conventional filters, membranes can be used in dead-end or cross-flow modes. However, unlike conventional filters most membranes are used in cross-flow. Although dead-end microfiltration is very important for the sterilization of gases and liquids, it is not suitable for clarification of fermentation broths because the filter elements rapidly become blocked by particulates. This text concentrates on cross-flow processes.

In clarification of fermenter broths the process stream is pressurized and passed tangentially across the surface of the membrane at flow velocities of up to 6 m/s. The shearing action of the tangential flow inhibits the build-up of solids on the membrane surface and helps to maintain good performance. Theoretical treatments of this process are generally rather poor for the purposes of predicting performance. Nevertheless some of the simple models help to illustrate important aspects of membrane behaviour. These will be described here to illustrate guidance on the best ways to use membranes.

5.2.1 Film theory

Ultrafiltration and microfiltration membranes are surface filters, that is, solute and solvent are separated at the membrane surface. Solvent and non-retained solutes pass through the membrane—retained (or rejected) components accumulate at the membrane surface, but can diffuse back into the bulk of the tangentially flowing medium. This results in a reversible increase in solute concentration at the membrane surface known as concentration polarization. At steady-state the rate of convective transport of solute to the membrane surface is exactly counterbalanced by diffusive transport away from the membrane surface (see *Figure 6*). For a completely rejected solute, this can be represented mathematically by the following relationship:

$$JC = -D\frac{dC}{dx} \qquad [10]$$

where C is solute concentration at a point (kg/m^3), J is the permeate flux (m/s), D is the solute diffusivity (m^2/s), and x is distance measured form the centre line of the flowing stream (m). Rearranging and integrating across a boundary layer of thickness δ with boundary conditions of $C = C_b$ (bulk concentration) at $x = 0$ and

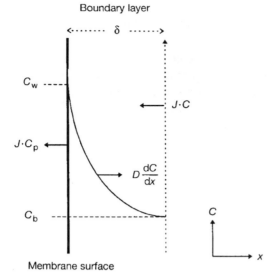

Figure 6 Concentration polarization at a membrane surface.

$C = C_w$ (concentration at the wall or membrane surface) at $x = \delta$ gives the following equation:

$$J = \frac{D}{\delta} \ln \frac{C_w}{C_b} \qquad [11]$$

Because of the near impossibility of measuring the thickness of the boundary layer, δ, this is more usually written:

$$J = k \cdot \ln \frac{C_w}{C_b} \qquad [12]$$

where k is a mass transfer coefficient equivalent to D/δ. Mass transfer coefficients can be estimated independently from a range of correlations. The following are often used but are not unique:

Laminar flow

$$k = 0.664 \left[\frac{u}{L}\right]^{1/2} \cdot \frac{D^{2/3}}{\left(\frac{\mu}{\rho}\right)^{1/6}} \quad \text{Grober} \qquad [13]$$

Turbulent flow

$$k = 0.023 \frac{\bar{u}^{0.8} D^{0.67}}{d^{0.2} \left(\frac{\mu}{\rho}\right)^{0.47}} \quad \text{Dittus, Boelter} \qquad [14]$$

where $\bar{u}$ is the mean fluid velocity (m/s), d is the hydraulic diameter of membrane element (m), μ is the fluid viscosity (Nm/s^2), and ρ is the fluid density (kg/m^3). Laminar flow correlations apply where the Reynolds number, defined as $\rho u d/\mu$, is less than 2000.

In general, the concentration of solute at the membrane surface must reach a maximum value defined either by a solubility limit or by closest packing of molecules. Equation 12 therefore predicts that once the wall concentration reaches this value permeate flux will reach a limit that is independent of pressure and this is indeed what is observed in most cases. The equation also suggests that the limiting flux is independent of membrane properties, which does not hold quite as well in practice. Nevertheless for a given feed stream, different membranes often give very similar performance even where the initial membrane properties are very different.

5.2.2 Operating characteristics

To get the best from membrane processes it is important to be aware of their behaviour during operation.

When tested using pure water most membranes give very high fluxes. Depending on the type of membrane this can vary from around 100–10 000 litres/m²/h. However, during actual filtration the flux is likely to be very much lower, usually in the range 40–200 litres/m²/h. *Figure 7* illustrates typical flux-time behaviour for a cross-flow membrane process. At the start of operation, the flux is high and in some cases may approach the pure water flux. However, a rapid decline occurs over the first 20–30 min of operation followed by a gradual, but persistent, fall. This reduction in flux is caused by a combination of concentration polarization and fouling (see below). Eventually long-term fouling causes the flux to fall to a point where cleaning is necessary to improve membrane performance. It is rarely possible to restore flux to its original level, but periodic cleaning enables good fluxes to be maintained over the entire membrane life, which in industrial applications is around six months to two years.

Permeate flux is influenced by two controllable operating parameters; cross-flow velocity and transmembrane pressure. At low pressure, flux is almost directly proportional to transmembrane pressure. As pressure increases, however, the flux increases more slowly and eventually reaches a limiting flux where changes in pressure no longer have any effect on flux (see *Figure 8*). This limiting flux creates a theoretical and practical ceiling to membrane performance. Increasing cross-flow velocity also increases flux, at any transmembrane pressure. Improving hydrodynamic conditions such as by raising cross-flow velocity is the only

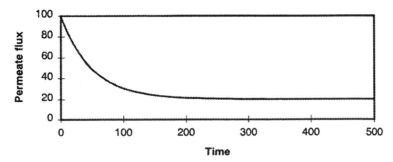

Figure 7 Typical flux decline curve for cross-flow membrane processes.

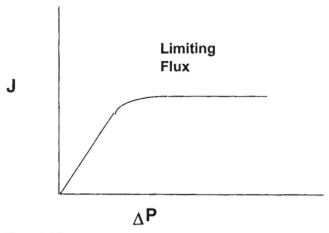

Figure 8 Effect of pressure on permeate flux.

way to increase the limiting flux, but again the effect diminishes at higher velocities.

5.2.3 Concentration polarization and fouling

The rapid initial flux decline described in the previous section is caused mostly by concentration polarization. Concentration polarization is the increase in solute or particulate concentration that occurs at a membrane surface as a result of the separation of solute and solvent. The raised particulate concentration increases the resistance to solvent flow and thus reduces flux. Concentration polarization is a reversible, but unavoidable phenomenon. It can be reduced by improving mass transfer within the membrane element, but can never be wholly eliminated.

Fouling is the generic name for a number of phenomena that cause irreversible changes in membrane properties and the effects of which can only be reversed, if at all, by cleaning. Fouling is responsible for at least part of the rapid initial drop in membrane performance and is the cause of long-term flux decline. In protein purification, the main modes of fouling are macromolecular adsorption, pore blockage, and cake formation.

i. Solute adsorption

Adsorption occurs due to a combination of hydrophobic and electrostatic interactions between macromolecules and the membrane surface. It is a spontaneous process that takes place on contact between the membrane and the stream to be filtered; no filtration is necessary. The presence of adsorbed material at the membrane surface causes an increase in resistance to flow. In the case of proteins, adsorption can result in up to a threefold reduction in flux depending on the type of membrane used. Adsorption can be minimized, but not wholly prevented, by using membrane materials with low non-specific binding. Tests for identifying suitable membrane materials are described in Section 5.4.

ii. Pore blockage

Pore blockage occurs where particles become lodged at pore entrances or adhere to the pore walls. The consequence is a reduction in the area available for permeate flow and a drop in flux. Often there is also an increase in the retention of small species. This is particularly significant where membranes are being used to separate a protein mixture from cells or cell debris. Reduction in pore radius can lead to significant retention of protein by membranes with pore sizes two orders of magnitude larger than the protein. Pore blockage is most severe when the ratio of particle to pore diameter is in the range 0.5–2.0. For this reason, provided that all dissolved and suspended solids are to be concentrated, it is often best to use membranes with a pore size very much smaller than the smallest species present.

iii. Cake formation

Concentration polarization leads to the formation of a concentrated layer close to the membrane surface. In principle this layer consists of discrete particles that will disperse once filtration stops, but in reality interparticle interactions result in aggregation and eventually in the formation of a consolidated cake layer that in effect creates a secondary membrane. This cake layer can cause retention of small particles and gradual build-up may be responsible for long-term fouling.

5.3 Membrane equipment

Membranes are available for a wide range of applications and this is reflected in the large variety of commercially available equipment. The main elements to consider are the membrane type and the design of the membrane housing or module.

5.3.1 Membrane types

The most important property of a membrane is the pore size. These range from reverse osmosis membranes which can separate salt from water to micro-filtration which employs membranes with pore sizes of between 0.1 and 5 μm. Although the separation characteristics of membranes represent a more or less continuous spectrum of possibilities between these extremes, there are distinctly different mechanisms of separation at either end of the scale. Microfiltration membranes act entirely by a physical sieving process, but as the pore size of the membrane decreases physicochemical interactions between the membrane and solvent or solute become more important. Reverse osmosis membranes, at the opposite extreme, are thought to be completely non-porous and act entirely by differential diffusion within the membrane phase.

The most important membrane types for clarification of fermentation broths are ultrafiltration and microfiltration, both of which act largely by a sieving mechanism. Although there is no definitive distinction between the two processes, ultrafiltration membranes are considered to have pore sizes of macromolecular dimensions (usually expressed as nominal molecular weight cut-off) and micro-filtration membranes have pore sizes of the same size as micro-organisms and are rated according to nominal pore diameter. In principle, microfiltration membranes with their larger pore sizes, better matched to the particle size, should be better for clarifying fermentation broths than ultrafiltration membranes. However, this is not always the case.

Most membranes are made from polymeric materials usually cast onto a porous support. The specific material affects range of possible pore sizes, mechanical strength, fouling behaviour, and resistance to attack by heat or chemicals. Despite their high cost, ceramic and metal membranes are used in a number of biological applications because of their ability to withstand repeated steam sterilization.

5.3.2 Membrane modules

On an industrial scale, membranes are supplied in the form of modules that can contain up to 50 m^2 of membrane area. These modules come in a variety of forms: hollow fibre, spiral wound, flat plate, and tubular. Each design is suited to particular applications and, particularly on a large scale, process economics are critically dependent on choosing the right module type. At laboratory scale the choice is often less critical, but nevertheless has an effect on performance.

The commonest forms of cross-flow membrane for laboratory operation are flat sheet and hollow fibre modules. Each has its own particular advantages. Hollow fibre membranes consist of a bundle of fibres 0.5–2.0 mm in diameter, which are potted into a cylindrical casing. They are very compact and accommodate a large membrane area in a small volume. They are also relatively inexpensive per unit of membrane area. However, a combination of low burst pressure and narrow feed channels, limits these membranes to low pressure operation with low viscosity feeds. Some solids can be tolerated, but hollow fibres are best suited to particle-free feeds. However, some manufacturers have developed hollow fibres with internal diameters of 1.5–3 mm, which are useful for clarification of cell suspensions.

Most of the major manufacturers (Millipore, Pall Filtron, Sartorius) produce flat sheet modules with membrane areas from a few square centimetres up to several square metres. The smaller units are often sealed, disposable units whereas larger devices have frames into which a series of replaceable membrane elements can be mounted. The flow channels in flat sheet membranes can be either open or screened. Screened channels have plastic meshes that both keep adjacent sheets of membrane apart and increase turbulence within the process stream. The screen increases mass transfer within the channel, reducing accumulation of deposits and increasing flux. Screened channels are not suitable for streams with high levels of particulate as particles become trapped in the mesh, ultimately leading to channel blockage. The wider, unobstructed flow channels in open channel flat plate system make them suitable for processing streams with a high viscosity or high solids concentration such as fermentation broths.

Where ceramic and metal membranes are available for laboratory scale operations, it is usually in the form of tubular membrane elements. These are similar in shape to hollow fibres except that the flow channels are typically much wider, in the range 3.0–12.5 mm. The main consequences are: that the units are much larger for a given membrane area and require larger ancillary pumps and pipework. Provided these are not an issue this form of module is well suited to clarification of fermentation broths.

One other form of membrane module to mention here is the stirred cell. These are strictly for small scale applications and can process volumes of up to about 1 litre. Stirred cells consist of a cylindrical vessel the base of which can be unscrewed and used to house a disc of membrane. The cap of the cylinder contains connections for application of compressed air and a spindle on which is mounted a magnetic stirrer bar. These are essentially dead-end filters, but the stirrer bar can be used to agitate the feed so as to achieve some of the benefits of tangential flow, namely reduced build-up of particulates at the membrane surface. They are also useful for measuring the filtration properties of membranes.

5.4 Membrane selection

One of the main advantages of membrane processes over conventional filtration is that they are more selective and thus capable of producing filtrates with fewer impurities. One consequence of this selectivity, however, is that membrane equipment must be chosen far more carefully to fit each application. Membranes are available for a wide range of applications and this is reflected in the large variety of commercially available equipment so making the correct choice is critical to successful operation. This section looks at factors to consider when choosing membranes and describes a few simple tests to help narrow the choice. There are three main aspects of membrane behaviour to consider: the membrane pore size (separation characteristics), membrane material, and module type.

5.4.1 Pore size

Pore size affects both the separation characteristics and the permeate flux. The influence on separation is obvious: particles or solutes larger than the pore size are retained at the membrane surface and smaller species pass through. In practice, there is not a sharp distinction in size between retained species and those that pass through the membrane. This is partly because, with a few exceptions, membranes do not possess pores of a single size. Instead there is a distribution of pore sizes and for a particular particle, rejection will depend on whether it approaches a small or a large pore. However, even if pores were of a uniform size, there still would not be a sharp cut-off as hydrodynamic (frictional) and steric (shape) effects would ensure that some particles smaller than the pore size would still be retained by the membrane.

The effect of pore size on permeate flux is somewhat more complex. Large pore size membranes usually have greater permeability owing to the greater ratio of pore area to membrane surface area and the lower ratio of pore surface area to pore volume. Both lead to a reduction in frictional interaction between membrane and solvent. However, when particulates or retained solutes are present, the flux can be 2–3 orders of magnitude lower than the pure solvent flux due to the effects of concentration polarization. For a given feed stream, there may be relatively little difference in flux produced by different membranes, in line with the predictions of the film theory (Section 5.2.1). Nevertheless, larger pore size membranes generally give slightly higher fluxes, provided that pore size is substantially less than the particle or solute size. However, where the ratio of particle diameter to pore size is close to 1 pore blockage can occur leading to a rapid reduction in flux.

Unless small molecules are to be eliminated in the permeate, choose a pore diameter no more than half the expected diameter of the smallest particle. Where small species are to be recovered in the permeate, pore size must be a balance; large enough to enable good transmission of small species, but small enough to prevent pore blockage.

5.4.2 Membrane material

There are three important considerations in the selection of membrane material. First, it should be available with the correct pore size and in an appropriate modular form. Secondly, it should be resistant to fouling by constituents of the feed, particularly the product. Finally, it must be resistant to chemical attack by components of the feed stream and any cleaning or sanitizing agents used to restore membrane performance after each operating cycle.

Information on pore size ranges and module design can be found from manufacturers' literature.

To some extent compatibility of membrane material with common chemicals can also be obtained from the manufacturer. However, information is unlikely to be definitive and some testing is recommended whenever aggressive chemicals are to be used.

The simplest form of test is to soak membrane samples in solutions of the chemical of interest and to measure the change in water flux (see *Protocol 3*). Provided the membrane material is compatible with the chemical, there should be little or no change in the water flux. A significant decrease suggests either binding to the membrane material or perhaps a partial collapse of the membrane structure. An increase in water flux, on the other hand, might indicate chemical attack on the membrane structure.

Protocol 3

Test for chemical compatibility of membrane materials

Equipment
- Sheet of membrane to test
- Stirred cell or small module which can be loaded with flat sheets of membrane
- Ancillary equipment for operation of membrane module

Method

1 Identify any chemicals in the feed or cleaning agents that might cause damage to the membrane.

2 Consult published charts of chemical resistance to determine whether the selected membranes are compatible. Where no information exists, testing will be necessary. Wash thoroughly to remove preservatives and wetting agents.

3 Cut out a disc of membrane to fit into a stirred cell or other dead-end filtration device.

4 Wash thoroughly to remove preservatives and wetting agents.

5 Load into cell and place pure (Analar grade or similar) water in the feed reservoir.

6 Apply a fixed pressure and measure the permeate flow rate. A pressure of around 1 bar will be appropriate in most cases, but this may need adjustment for very permeable or low permeability membranes.

7 Soak membrane in a solution of the suspect reagent (test at a range of concentrations and exposure times if necessary).

8 Rinse the membrane with Analar grade water and repeat steps 5 and 6.

9 Repeat procedure for each reagent.

Adsorption of macromolecules and particulate material can cause severe membrane fouling and loss in performance. One of the best ways to minimize the effect is by choosing membranes to which feed stream components do not bind. In principle it is possible to measure the quantity of material that a membrane binds and this has been found to be an excellent indicator of fouling problems. However, this approach requires specialist equipment and is expensive. *Protocol 4* provides a simpler method, but which produces adequate results for small scale devices.

Ideally the water flux should be the same before and after the run. Where the feed is a pure protein solution the change in flux will be negligible with low protein binding membranes, but can be very high when the membranes exhibit appreciable non-specific binding. For more complex feeds, such as fermenter broths there will always be a slight flux decline, but the most suitable membrane types will be those that give the smallest change.

Protocol 4

Assessment of membrane fouling properties

Equipment

• See *Protocol 3*

Method

1 Cut out a disc of membrane to fit into a stirred cell or other dead-end filtration device.

Protocol 4 continued

2 Wash thoroughly to remove preservatives and wetting agents.

3 Load into cell and place pure (Analar grade or similar) water in the feed reservoir.

4 Apply a fixed pressure and measure the permeate flow rate. A pressure of around 1 bar will be appropriate in most cases, but this may need adjustment for very permeable or low permeability membranes.

5 Replace water with a sample of feed and leave in contact with the membrane for 1 h. No pressure should be applied and the permeate line should be closed to prevent permeation of liquid.

6 Remove feed and rinse both membrane and module with pure water.

7 Place pure water in the feed reservoir and measure the permeate flow rate under the same conditions as before.

8 Repeat for all candidate membranes.

5.4.3 Membrane module

Membrane modules can be divided into two broad categories; wide channel and narrow channel. Narrow channel devices (spiral wound, hollow fibre, and screened channel flat plates) are the cheapest and therefore preferred wherever possible. However, these designs are not suitable for feeds, such as fermentation broths, that contain high levels of solids as the channels are prone to blockage. In such cases, it is necessary to use wide channel systems (tubular, open channel flat plates, and large bore hollow fibre).

Once a suitable membrane has been identified from the above procedure, the next step is to determine appropriate operating conditions. A method for achieving this is described in the next section.

5.5 Membrane operation

One of the characteristic features of cross-flow membrane operation is that permeate flow rate is very much lower than the retentate flow (flow across the membrane surface); typically 1–5% depending on module type. Since in most cases the aim is to recover close to 100% of liquid in the permeate it is necessary to recycle the feed. Consequently, almost all cross-flow membrane processes are operated as a recirculating loop.

5.5.1 Batch operation

The simplest set-up, often called open-loop batch, is depicted in *Figure 9*. A single pump is used to circulate liquid from a feed reservoir through the membrane module and back to the feed reservoir. In the simplest version all of the feed is placed in the reservoir initially and operation is stopped when either a set volume or concentration is reached in the feed reservoir. A slightly more sophisticated version is the fed batch system where additional feed is added continuously to the reservoir. Again operation is normally terminated after a

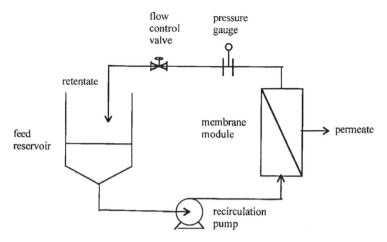

Figure 9 Open-loop batch operation.

particular volume has been processed or when a target concentration has been reached.

A third form of batch system, closed-loop batch operation, is shown in *Figure 10*. This type of plant has a high pressure loop in which liquid is circulated at pressure by a centrifugal pump while a smaller positive displacement pump is used to feed liquid into the system at the same rate as permeate is withdrawn. One advantage is that it avoids some of the energy losses of an open batch system where feed is throttled back to atmospheric conditions at each cycle. It is, therefore, well suited to processes, such as reverse osmosis, that operate at high pressure. Another benefit is that it enables membranes to be operated at constant flux (see below) and gives greater flexibility in setting operating conditions to minimize the effects of fouling. The main disadvantages are increased complexity and cost.

One important effect of recycle operation used in membrane processing is heating of the feed stream. Where there are thermally labile products or where it is essential to prevent bacterial growth, cooling coils should be installed in the membrane loop to control temperature.

5.5.2 Diafiltration

Operation of membranes as described in the above section produces a clarified permeate containing proteins and other low molecular weight impurities, and a concentrated stream containing cells or cell debris plus the same dissolved species (proteins plus impurities) as the permeate. Typically no more than 70% of the protein is recovered in the permeate stream and sometimes less. Where higher recoveries are required the standard clarification step should be followed by a diafiltration stage. This can be achieved quite simply by adding water to the feed reservoir during operation, either in batches or to match the rate of permeate flow. The effect is to flush dissolved species into the permeate. The

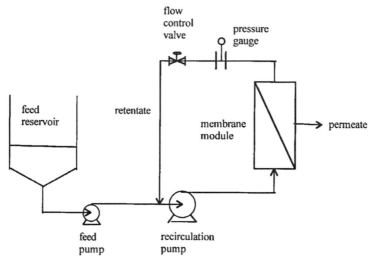

Figure 10 Closed-loop batch operation.

process should be continued until the required amount of clarified product or protein recovery has been achieved.

5.5.3 Constant pressure operation

Membrane processes are most commonly operated at constant pressure, i.e. the pump and control valves are set to deliver a constant pressure and a fixed re-circulation flow rate. No attempt is made to control flux through the membrane. The result is that at the beginning of a cycle the flux is very high, falling rapidly over the first 30 min of operation. Subsequently, flux is usually fairly steady, but does continue to decline gradually.

This method has the advantage of being extremely simple and in most cases gives satisfactory performance. However, overall efficiency may be substantially reduced due to high fouling rates at the start of the cycle when the flux is highest.

5.5.4 Constant flux operation

The performance of membranes used in constant pressure operation declines continually due to the effects of fouling. Some researchers have noted that, for some applications, membranes operated below a certain flux experience no fouling at all. In other cases, it has been observed that reducing the flux at the beginning of a run can significantly reduce fouling and enhance productivity over the duration of a run. This reduction in fouling can be achieved by using constant flux operation, which can be produced either by using a closed loop system in which feed is injected into the loop by a positive displacement pump delivering a constant flow rate or by placing a flow controller on the permeate line. For most simple laboratory operations the additional complexity will prob-ably not be justified, but for long-term operation considerable improvements may be possible.

5.5.5 Operating conditions

Operating conditions have a significant effect on membrane performance, both the flux (treatment rate) and protein transmission (recovery). For a given membrane module the main parameters that can be set are the transmembrane pressure and the cross-flow velocity (recirculation rate). Protein recovery is generally best at high cross-flow velocity and low pressure; flux is usually highest at high cross-flow velocity and high pressure. It is, therefore, usual to operate at or close to the highest cross-flow velocity permitted by the membrane module, but the best pressure must be a compromise. As was illustrated in Section 5.2.2, permeate flux reaches a limiting value as pressure is increased. In most cases, the aim should be to operate at the point where the limiting flux is just reached. One method of finding this point is described in *Protocol 5*.

Protocol 5

Determination of limiting flux

Equipment and reagents

- Membrane unit as shown in *Figure 9*
- Timer
- Sample of feed

Method

1. Open flow control valve fully and start recirculation pump at low speed.
2. Adjust feed flow rate to desired value and start timer.
3. Measure permeate flux at 5 min intervals for 20 min.
4. Slowly close flow control valve until the mean transmembrane pressure has increased by 0.1 bar; adjust pump speed to maintain a fixed recirculation flow.
5. Measure permeate flux at 5 min intervals for 15 min.
6. Repeat steps 4 and 5 until there is no further increase in flux (or the maximum recommended operating pressure for the membrane module has been reached).
7. The final value is the limiting flux; operate at the penultimate pressure tested except where the maximum operating pressure has been reached.

In summary, membranes should be operated at high cross-flow velocity, and the pressure should never be exceeded the value at which limiting flux is just reached.

5.5.6 Membrane cleaning

Chemical cleaning is an essential part of membrane operation, partly to restore flux and, in biotechnology applications, to prevent microbial contamination. A variety of chemicals can be used to clean membranes. For protein recovery the most common cleaning agents are combinations of sodium hydroxide and detergents and these can either be laboratory reagents or commercial brands.

A typical membrane cleaning regime is described in *Protocol 6*. This method can be used both during routine operation and to determine the most effective cleaning chemicals. The most important parameter is the ratio of water flux after cleaning to that of the clean membrane prior to the previous run. Ideally this ratio should be close to 1.0. There is cause for concern if the ratio is greater or significantly less than 1. If it is greater than 1 the cleaning regime may be causing damage to the membrane. Where the ratio is much less than 1, it may be difficult to clean the membrane and performance may rapidly become unacceptably poor. The first time a membrane is used it may be difficult to restore water flux to more than about 80% of its original value. This is not unusual and should not be regarded as a problem with the cleaning regime. However, cleaning after subsequent cycles should restore water flux to roughly constant level.

Protocol 6

Membrane cleaning

Equipment and reagents

• See *Protocol 5*

Method

1 Complete a normal operating cycle.

2 Drain membrane system.

3 Flush with water for around 5 min and drain.

4 Measure water flux.

5 Fill system with cleaning solution and circulate at moderate temperature (usually around 50 °C) and at as low a pressure as possible for 5–15 min. Note cleaning solution should not be circulated for too long as material removed from the membrane surface can start to re-deposit and may well be more difficult to remove the second time around.

6 Drain system, rinse with water, and measure water flux.

7 Repeat steps 4 and 5 until water flux approaches the value prior to the operating cycle. If the flux cannot be restored to this level, cleaning should be stopped when the washing cycle has no further effect.

6 Choice of clarification process

Each of the processes described in this chapter has particular advantages and disadvantages and these are summarized below.

The main advantage of conventional filtration is the simplicity and ease of use. It is very convenient for laboratory operations up to about 5 litres in volume. Larger volumes may exceed the capacity of the simple systems described in this chapter and necessitate the use of more sophisticated equipment. Containment of the process is poor which could be problematical for hazardous streams. In general, the filtered product is likely to contain small quantities of cellular material, so the process may not be suitable where subsequent processes are sensitive to solids. Without washing of filtered solids there will be some loss of protein products.

Of all the processes described, membrane filtration gives the highest clarity product, i.e. the lowest content of cell debris. However, protein losses may be relatively high. High protein recovery can be achieved using diafiltration, but this does result in some dilution of the product. Membrane processes are relatively complex even at small scales of operation. For volumes less than 5 litres it is better to use conventional filtration unless filtrate clarity is an issue. Membrane processes are best suited to intermediate volumes of 5–100 litres.

Small scale laboratory centrifuges give good protein recovery and in most cases good product clarity. Typical swing or fixed rotor machines are ideal for clarifying small quantities of feed up to around 500 ml depending on the capacity of the particular design. Continuous flow machines, such as disc stack centrifuges, generally give good product recovery and clarity though not as good as small scale devices. These machines are expensive but are the most reliable process for larger scale separations. Continuous centrifugation is the preferred process for volumes over 50–100 litres.

Chapter 5

Cell disintegration and extraction techniques

R. H. Cumming and G. Iceton

School of Science and Technology, University of Teesside, Middlesborough TS1 3BA, UK.

1 Overview

Disruption of cells is necessary when the desired product is intracellular. Many commercial intracellular products are proteins, such as soluble enzymes or soluble genetically engineered peptides. However, a number of genetically engineered proteins in *E. coli* are present in the cell as insoluble inclusion bodies; although this may be due to the chosen lysing conditions (1). This chapter concentrates on the extraction of soluble enzymes.

To achieve a good yield of an intracellular product, it is generally a good idea to minimize the number of steps involved in the purification (2), as there is loss of material associated with each step. Since cell disruption is an extra procedure which itself may demand a further clarification step, it may be worth investigating if extraction of the protein can be made directly from the cell lysate (the disrupted material) without a cell debris clarification step. It may also be possible to manipulate the cell so that it excretes the protein into the surrounding growth medium, and thus, not require an extraction procedure (if the subsequent dilution of the product by the broth can be tolerated).

The choice of disruption method usually follows one of the two directions:

(a) Can a given disruptor be used for a particular cell type?

(b) Which is the best method of extracting a product?

The former often occurs in a laboratory context, whilst the latter is a question to be asked during scale-up of a process when costs are paramount. This chapter tries to accommodate both questions.

Whatever type of disruption process is adopted, there are some key questions to be addressed. These are briefly discussed in the remaining sections of this overview. A recent comprehensive review of many disruption techniques is given by Middelberg (3).

1.1 Stability of the released protein

It is well known that some enzymes are more stable than others. The disruption method can impose great physical and chemical stress on the enzyme. Enzymes which are stable in the cell, perhaps by virtue of being attached to membranes, will be released into the medium during disruption. Since the chemical environment of a disrupted cell is very different to that of an intact cell, careful attention to the composition of the buffers used for the disruption process can minimize damage to the product. A number of the mechanical disruption processes will expose the enzyme to high shear stresses, which in combination with other factors, such as gas bubbles, have been shown to inactivate enzymes. Many of the mechanical disruption processes also produce considerable heat, which can result in denaturation of the enzyme.

Proteolytic enzymes which are naturally occurring in all cells may be released into the lysate during disruption and can degrade the target enzyme. To minimize this, the disruption process is often carried out at low temperatures which slows proteolytic activity. Additionally, specific or general protease inhibitors can be included in the disruption buffer.

Extraction of enzymes from plant tissues often presents the researcher with an additional problem in the form of phenolic compounds. Phenolic compounds bind to proteins and will inactivate enzymes. Unfortunately, during the disruption process both the desired product and phenolic compounds will be released from the normal location in the cell. Sometimes, it may be necessary to include polyphenol-binding agents in the homogenization buffer to minimize this threat. If substantial damage to the protein product occurs during the disruption process, the yield will be low.

1.2 Location of target protein within the cell

The target enzyme will usually be contaminated with other enzymes and structural proteins released from the cell during disruption. This makes separation of the product from the mixture more difficult. Efforts have been made to selectively release enzymes during a disruption process. It would be expected that during a mechanical disruption method, cytoplasmic enzymes would be released before membrane-bound enzymes, since extra energy is required to dislodge the enzymes from the membrane. A number of workers have reported a sequence of enzyme release during disruption (e.g. ref. 4), but this has been quite difficult to exploit.

Some success in selective release of intracellular products has been claimed by Chi *et al.* (5), in the extraction of hepatitis B virus surface antigen (HBsAg) as peptide particles from recombinant *Saccharomyces cerevisiae*. Because the HBsAg particles were associated with the cell membrane of the yeast, the workers were able to devise a two-step procedure for HBsAg extraction which gave a better yield than a single-step procedure. The two-step procedure used a bead mill disruption stage to release the soluble protease, keeping the HBsAg attached to membranes associated with the cell wall fraction. The cell wall debris was then isolated by centrifugation and extracted with a non-ionic detergent Triton X-100. In this way, the product was associated with much fewer contaminating proteins than in the one-step process which used a bead mill with Triton X-100 in the extraction buffer. Another advantage of the two-step process was that proteases were removed during the first extraction, which helped to minimize product degradation. However, other workers have found that solubilization of cell wall bound enzymes precedes the release of cytoplasmic enzymes. Thus, the release of invertase from the cell walls of yeast can be achieved without extensive destruction of the cell (6).

1.3 The yield and kinetics of the process

The *yield* of product achieved in a disruption process is of much greater importance in a large scale process than in a laboratory one. The *yield* is the quantity of enzyme released per unit starting material. It will have units of amount of enzyme per gram, for instance. It may be useful to express this in terms of specific activity as the amount of enzyme per gram of protein released. The amount of enzyme will be expressed in units of enzyme activity. If a general investigation into the disruption process is being carried out, then the extent of disruption can be measured in terms of the total intracellular protein released, rather than the specific enzyme.

It will be found that the yield of enzyme from cells will depend on the disruption process. Factors which contribute to the differences are:

(a) Location of product within the cell.

(b) Degree of disintegration.

(c) Extent of denaturation of the product during disruption.

The *rate* of a disruption process is important too, since it is possible to stop the disruption process before it is complete, which thus, results in a low yield. It is often desirable to disrupt the cells until the *maximum yield* has been reached. For mechanical processes the rate can be modelled by a first-order process. For a time-dependent process such as bead milling or ultrasonication the rate is given by Equation 1, where R, R_M are the released protein concentration and the maximum released protein concentration respectively; k is the first-order rate constant, and t is the disruption time:

$$R = R_M (1 - e^{-kt}) \qquad [1]$$

If the disruption process is based on the passage of the cells through a disruption device such as in homogenization, then the rate Equation 1 has to be modified such that t is replaced by N, the number of passes through the device:

$$R = R_M (1 - e^{-kN}) \qquad [2]$$

Both Equations 1 and 2 produce characteristic curves (*Figure 1*). It is, thus, evident that once about 95% disruption is achieved, improvements in yield by extending the disruption time or number of passes will be small. Further, it can

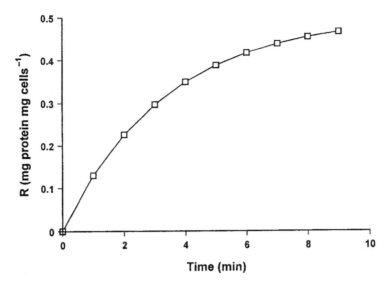

Figure 1 The shape of the release curve for first-order kinetics (Equation 1), with R_M = 0.5 mg protein/mg cells and k = 0.3 min^{-1}.

be detrimental to expose the cells to prolonged disruption treatment since denaturation of an enzymic product can occur. This leads to the concept of an *optimum* disruption time. A number of research groups have modelled the disruption process in terms of an optimum (e.g. ref. 7).

Non-mechanical disruption methods such as chemical and enzyme treatment are also time-dependent, although their rates have been less extensively researched. Enzyme digestion of yeast cells is quite slow, as can be solvent extraction. Detergent treatment is fast.

Since the optimum time for extraction will depend on the biological components (cell type and enzyme) as well as the method of extraction, it is evident that the time of extraction given in the following protocols will be approximate; and could usefully be changed if the yield is low.

1.4 Continuous or batch disruption

If large quantities of cells need disrupting, it may be useful to consider a continuous disruption process. Both homogenization and bead-milling have been successfully employed on a continuous basis. In the case of a bead mill, the extent of disruption can be calculated by a mass balance to give:

$$R = R_M \frac{kV}{F + kV}$$
[3]

where R, R_M are the protein yield in the supernatant and maximum protein yield possible (e.g. mg/ml or mg protein/mg cells^3), F is the flow rate through the mill (cm/min), V is the volume of suspension in the mill (ml), and k is the first order rate constant (min^{-1}). It can be seen that increasing the flow rate decreases R, the protein concentration in the exit stream.

The homogenizer is often in a recycle mode and a fraction of the disrupted material is returned to the homogenizer input line. This is because one pass of the material through the homogenizer is seldom sufficient to achieve a satisfactory protein yield.

1.5 The need to consider subsequent steps

The disruption step should not be considered in isolation from the subsequent protein purification steps. Usually the size of the particles (cells) will be reduced as disruption occurs. Additionally, nucleic acid will be released from the cell which can increase the viscosity of the lysate enormously. The cell debris generated has to be removed in a clarification step (unless a purification step such as liquid–liquid extraction is used). The physical properties of the lysate (small particle size and high viscosity) make any clarification step difficult. The nucleic acid can be hydrolysed by the addition of a nuclease or other agents, e.g. precipitants: polyethyleneimine or polyethylene glycol (8); or by heat treatment (9). The clarification step is usually centrifugal, since the small particle size clogs membranes filtrates, and *in the laboratory* centrifuges which reach high g are available. On a larger scale, the availability of a suitable debris removal stage (e.g. high g centrifuge) can influence the choice of disruption process. An approximate check as to whether a laboratory centrifuge is capable of sedimenting the cell debris produced during mechanical disruption of cells can be made by using the equation:

$$t = \frac{\ln (R_S / R_L) \times 18\mu}{\omega^2 (\rho_P - \rho_L) d_p^2} \qquad [4]$$

where t is the centrifugation time, R_S, R_L are the radii from the centre of rotation of the centrifuge head to the liquid surface and bottom of the centrifuge tube, respectively. The liquid viscosity is given by μ, while ρ_P and ρ_L are the particle and liquid densities, respectively. The angular velocity, ω, is calculated from the centrifugation speed (in r.p.m.) by the following equation:

$$\omega = \frac{2\pi \, (\text{r.p.m.})}{60} \qquad [5]$$

An estimate of the mean diameter of the cell debris is given by d_P. SI units should be used. Typical values for (yeast) cell debris assuming a mean particle size of 0.3 μm, are $\mu = 1.02 \times 10^{-3}$ Pa s, $\rho_P = 1.04 \times 10^3$ kg m^{-3}, $\rho_L = 1.02 \times 10^3$ kg m^{-3} (some data in ref. 10). Thus, a typical bench centrifuge requires at least 12 min at 15 000 r.p.m. ($R_S = 0.15$ m and $R_L = 0.1$ m). A cooled centrifuge is highly desirable for this length of time. Failure in obtaining complete separation of debris and supernatant may result in an overestimation of extracted enzyme, since enzyme incompletely removed from cell wall fragments would still be present in suspension.

The difficulty of clarifying cell lysate on a large scale has been investigated by Clarkson *et al.* (11) The cells were disrupted by homogenization with a pilot scale homogenizer (K3, APV) and collected in a Westfalia disc stack centrifuge

(SAOOH 205, Westfalia Milton Keynes, UK) at 10 000 r.p.m. (8470 g). It was found that after five passes of yeast cells through the homogenizer, only 67% of the particles were removed. Flocculation of the debris with borax allowed up to 85% of the debris to be collected in a decanter centrifuge.

Consideration must also be given to the effects on downstream processing of adding lytic agents to the cells. For example, will their presence in the cell lysate interfere with subsequent purification steps, such as chromatography? This is a distinct advantage of mechanical disruption methods where no chemicals need to be added.

1.6 Assessing the extent of disruption

If it is convenient and easy to assay for the desired protein product during a disruption process, then there is no need for an alternative method of assessing the degree of cell disruption. However, it is often more convenient to use a marker substance to assess intracellular protein release. This is usually an assay for total protein. Assay of the released protein is also useful from a different perspective; it allows the calculation of specific activity for a release curve. If the maximum specific activity of an enzyme product is associated with a particular time of disruption treatment, the disruption process could be optimized for this time. A decrease in specific activity during disruption can occur through the release of non-product protein which dilutes the enzyme. Thus, if target protein is cytoplasmic, release from the cell could occur under relatively mild disruption conditions. Further cell disruption would extract membrane-bound proteins, which dilute the target proteins. Generally, it is better to have cell extracts of high specific activity for any subsequent purification steps.

Another explanation for a fall in specific activity with disruption time is that enzyme denaturation is occurring during the disruption process. This has been modelled by a number of workers. One of the simplest models is that by Augenstein *et al.* (7) which allows the prediction of the concentration of active enzyme (E) to be predicted at any disruption time (t):

$$E = A \left(1 - e^{-k_1 t}\right) B e^{-k_2 t} \tag{6}$$

where k_1 and k_2 are the first-order rate constants for protein release and loss of specific activity, respectively. A and B are experimentally determined constants based on the maximum releasable protein and specific activity, respectively.

Protocol 1

Estimation of total intracellular protein[a]

Equipment and reagents
- Boiling water-bath
- 4 M NaOH
- Protein assay solutions (e.g. as for Lowry)
- Cell suspension

Protocol 1 continued

Method

1   Add 0.5 ml of 4 M NaOH to 0.4 ml of washed cell suspension (about 100 µg of cells).

2   Incubate cells at 100 °C for 5 min (treat standard protein solutions in the same way).

3   Cool and assay for protein by Lowry or Biuret method (dilute if necessary).

[a] Based on ref. 12.

The total intracellular protein (R_M in Equation 1) is often determined by completely extracting the protein from the cells with sodium hydroxide (*Protocol 1*) . However, this may not correspond to the R_M of the actual disruption process, hence, $R/R_M =$ would never reach 1. Thus, many workers consider it better to use the R_M actually reached in the disruption process.

1.7 Marker substances for cell disruption

If the actual product is not easily determined in a cell lysate, *marker* techniques are used. Reasons for not being able to assay for the product during disruption may be that its concentration is too small or there is no rapid and convenient assay.

Marker techniques can be categorized into three groups:

- biological
- physical
- chemical

A viable cell count is an example of a biological technique. It is reasonable to assume that as cells are broken in the disruption process, their viability decreases. In addition to being rather tedious, this assay does not indicate the intracellular materials have been released from the cells; merely that cell death has occurred.

Physical measurements are usually quick and non-invasive. One of the easiest methods is by measuring the volume of intact cells after a standard centrifugation period. Davies (13), in a study of the ultrasonication of yeast cells, centrifuged the lysates for 20 min at 1000 g. He found it easy to distinguish between the intact cells which were concentrated in the bottom layer and cell debris layers above.

Simple optical density (OD) measurements can be used. The OD has been found to decrease when disruption takes place (14). This is probably due to decreases in particle size and a change in refractive index of the lysate associated with release of intracellular solutes. Viscosity measurements can also be used to gauge the progress of disruption, since release of the cell contents into buffer solution increases the viscosity of the lysate. This is particularly so if the nucleic acid released from the cells remains intact (as with chemical disruption methods). The lysate will become very viscous and noticeably shear thinning.

The most common chemical method for assessing cell disruption is to measure the protein concentration in the lysate during a disruption process. Some care must be given to the choice of protein assay. The Biuret method has been frequently used when the expected protein concentration is high; and the Lowry and Bradford methods for the detection of low concentrations. The Lowry method is of course, prone to interference from a wide range of substances. The protein assay must be set up to match the expected protein concentration in the lysate. An example of the approximation of protein concentration of a cell lysate prepared from a wet cell paste can be estimated as follows (using appropriate data):

(a) Concentration of wet cell paste in lysis buffer: 400 g/litre.

(b) Assumed water content of cell paste: 80%.

Therefore, concentration of cells as dry weight:

$$\frac{20}{100} \times 400 = 80 \text{ g/litre}$$

Assumed protein content of cells: 30% on a dry weight basis. Therefore, expected maximum protein content for cell lysate, assuming 100% protein release is:

$$\frac{30}{100} \times 80 \text{ g/litre} = 24 \text{ g/litre or } 24 \text{ mg/ml}$$

The Biuret protein assay has a typical range of 0–10 mg/cm, and so could be used with this cell concentration; with suitable dilution.

When concentrated cell suspensions are used, then it may be necessary to employ a correction for the change in aqueous volume during disruption (15), although most workers have ignored this correction factor.

It should be noted that the release of a specific enzyme may not follow the release of the total protein. For example Kuboi et al. (14) compared four methods of assessing disruption during ultrasonication: OD change, change in light scattering properties (related to particle size), protein release, and release of marker enzymes. A number of marker enzymes were released at a faster rate than that of the total soluble protein. Thus, a marker enzyme for the periplasmic space (acid phosphatase; EC 3.1.3.2) was released faster than cytoplasmic enzymes (glucose-6-phosphate dehydrogenase, EC 1.1.1.49). Selective release of specific enzymes during homogenization was first reported by Follows et al. (4).

1.8 Containment of the process

It may be necessary to contain the disruption process to avoid release of hazardous intracellular microbial products into the environment. A hazardous product could be due to the product itself (e.g. a genetically engineered hormone), or cell wall products, or enzymes from the cells themselves. Laboratory worker infection from E. coli disruption has been recorded (16). Of course, sometimes the release of an intact organism (e.g. genetically engineered) is not desirable.

Mechanical disruption methods are most likely to leak and attention has been given to design aspects for containment (17). Chemical disruption methods have an advantage here, in that the disruption can be performed in a sealed vessel.

1.9 Scale-up

Not all disruption processes will be suitable for large scale operation. Each of the mechanical disruption devices has its own problems such as keeping the power input/volume the same as a laboratory scale run; or simply the cost of the power input needed. Operating on a continuous scale keeps the reactor volume small. Homogenization seems to have been most successfully scaled-up, with large volume machines available. Chemical and enzymic methods are simpler to scale-up, requiring only a suitable mixing device for the introduction of the chemical. However, on large scale problems encountered may be due to the rise in viscosity of the lysate as DNA is released intact from the cells. The cost of the lytic chemical would be a major consideration in scale-up of a chemical process. It has been noted that cell disruption by autolysis would be volume-independent; which means that scale-up of this disruption technique would not present any difficulties associated with large volumes (18).

2 Methods of disruption

This section describes a number of the useful methods of cell disruption. It is focused largely on microbial cells, although some methods are indicated for animal and plant tissue. It should be noted that special procedures may be needed to conserve the biological activity of an enzyme during extraction which are generally applicable. Thus, *low temperatures* (e.g. 4°C) should be used for extraction procedures unless otherwise mentioned. Low temperatures both slow the rate of enzyme damage caused by proteases released from the cells as well as prevent thermal denaturation. In addition, *protease inhibitors* are often included in the buffers (e.g. 0.1 mM phenylmethylsulfonyl fluoride, PMSF); especially with eukaryotic cells.

Additionally, it may be necessary to destroy or remove the nucleic acid (especially DNA) in the lysate which produces a very viscous material, making separation of the cell debris difficult.

2.1 Pre-treatment of material

Many pre-treatments of cells for a particular disruption method are often combinations of two (or more) methods of disruption. Thus, a pre-treatment of freezing and thawing before a mechanical method of disruption has been found to increase the degree of cellular disruption. Note that a freeze–thaw step is often unwittingly included in another disruption procedure when the cells are stored at −20°C before disintegration. A preliminary pre-treatment of the cells with a solvent or detergent will aid release of many products during subsequent mechanical disruption. Examples are given in the appropriate sections.

2.2 General procedure notes

Disruption methods can be usefully grouped into two main categories shown in *Table 1*, according to their equipment requirements. Non-mechanical methods (group 2 in *Table 1*) have the advantage over mechanical cell disruption techniques in that the cells are not extensively disintegrated. This makes subsequent cell debris separation easier as the particle size is still large; perhaps that of the original cell dimensions. On the other hand, addition of extra chemicals to the system may interfere with downstream processing.

There are many operating variables for cell disruption methods. If cells are not lysing under the conditions employed, perform a release curve to check that the incubation time is sufficient to release the desired product. If there is still no success, try changing the operational parameters discussed for each protocol. In summary these are:

- concentrations of lysing chemicals (enzymes, detergents, etc.)

- bead size, concentration, and agitator speed for bead mills

- operating pressure for homogenization

- volume of sample for ultrasonication

Table 1 The main cell disruption methods

1. Methods needing specialist equipment
(a) Mixers and blenders
(b) Coarse grinding with a pestle and mortar
(c) Fine grinding in a bead mill
(d) Homogenization
(e) Ultrasonication
2. Methods using non-specialist equipment
(a) Freezing and thawing
(b) Osmotic shock
(c) Chaotropic agents
(d) Detergents
(e) Solvents
(f) Enzyme lysis

2.3 Mixers and blenders

These devices can only coarsely grind cells. They are not suitable for microbial cells but are useful for animal and plant tissues. A technique for plant cells is given in *Protocol 2*. This protocol employs a buffer containing many inhibitors and a reducing environment to stop denaturation of the released enzymes. Additionally, the plant tissue is likely to contain phenolic materials which will inhibit the Lowry protein- and dye-binding assays. A great wealth of detail is given by Gegenheimer (19), which should be consulted for the extraction of other types of plant tissue.

Protocol 2

Use of a blender for the disruption of plant tissues: non-fibrous leaves[a]

Equipment and reagents

- Waring blender
- Miracloth (Calbiochem): moist
- Cheesecloth: moist
- Homogenization buffer: 50 mM Tris–HCl (pH 8.0 at 25°C), 5% glycerol, monovalent cation (0.1 M KCl or 0.05–0.2 M $(NH_4)_2SO_4$), 1 × inhibitor mix
- Inhibitor mix: protease inhibitors plus reductants

- Protease inhibitors: 1 mM PMSF,[b] 1 mM benzamide,[c] 1 mM benzamidine–HCl,[d] 5 mM ε-amino-n caproic acid,[e] 10 mM EGTA,[f] 1 μg/ml antipain, 1 μg/ml leupeptin, 0.1 mg/ml pepstatin
- Reductants: 5 mM DDT,[g] 20 mM sodium diethyl dithiocarbamate; antiphenolic: 1.5% PVPP[h]

Method

1 Wash the leaves, and de-rib if necessary.

2 Pack the leaves into the Waring blender and add ice-cold homogenization buffer (1–2 litre per 1 kg leaves).

3 Grind for a few 5 sec bursts, packing the material as needed with the spatula.

4 Grind for a further 90–180 sec in 30 sec bursts.

5 Filter the homogenate through six layers of cheesecloth.

6 Pass the filtrate through two layers of Miracloth.

7 Centrifuge at 12 000 g for 20 min at 4°C to remove fine debris (this step can be omitted when larger volumes are used with precipitation by ammonium sulphate as the next step).

[a] From ref. 19.

[b] PMSF (phenylmethylsulfonyl fluoride): prepare 100 mM (= 100 ×) stock in dry 2-propanol or anhydrous ethanol (95% ethanol, 5% 2-propanol). Store at room temperature for up to six months.

[c] Benzamide: 0.1 M stock in ethanol. Can be prepared with the PMSF.

[d] Benzamide–HCl: prepare in aqueous 0.1 M stock.

[e] ε-amino-n caproic acid: prepare 0.5 M stock in water. Can be combined with benzamidine stock.

[f] EGTA (ethylene glycol bis (β-aminoethyl ether) N,N,N',N'-tetraacetic acid). For 0.5 M stock: suspend 0.05 mol in 80 ml water, titrate to pH 8.0 with concentrated or solid NaOH, and dilute to 100 ml.

[g] DTT (dithiothreitol): prepare as 1 M stock in deionized water. Store at −20°C.

[h] PVPP (polyvinylpolypyrrolidone): add to buffer and allow 2–24 h for full hydration.

2.4 Coarse grinding: pestle and mortar

The release of enzymes from fungal tissue can be achieved by grinding frozen mycelium in a simple pestle and mortar (*Protocol 3*).

Protocol 3

Disruption of fungal tissue by grinding in a pestle and mortar

Equipment and reagents

- Pestle and mortar
- Liquid nitrogen
- Lowry protein assay reagents

- Homogenization buffer: 50 mM Tris–HCl pH 8.0, 5% glycerol, 0.1 M KCl

Method

1 Chop 100 g fresh mushroom cups into 1 cm cubes, or divide mycelial culture into similarly sized clumps.

2 Place mycelial chunks into a pre-chilled mortar and cover with liquid nitrogen.

3 After a few seconds, when the mycelium should be frozen solid, grind the mycelium to a fine powder, topping up with liquid nitrogen as necessary to maintain the frozen state.

4 Allow to warm to about 4°C, then add 100 ml cold homogenization buffer, and agitate gently with a glass rod.

5 Filter the suspension through a coarse filter to remove debris then centrifuge at 12 000 g for 15 min at 4°C.

6 Assay supernatant for protein by the Lowry method (a 1:10 sample dilution may be necessary).

2.5 Fine grinding: the bead mill

Various forms of bead mill have been used for the disruption of micro-organisms. From the original work using yeast in a general purpose bead mill, mill design has been improved to increase the efficiency of disruption of bacteria and yeast. Milling is also an effective method for disrupting larger cells of algae and filamentous fungi (*Table 2*). Generally, the smaller the cell the more difficult its for the bead mill to break the cells.

Bead mills consist of a chamber in which beads and the cells are mixed in a suitable lysis buffer and agitated at high speed by a set of blades rotated within the chamber. The bead mills will have cooling jackets through which sub-zero coolant can be circulated, since considerable heat is developed during the disruption process. The cells are broken by being ground between the fast

Table 2 Conditions for bead milling different types of cells

Cells	Bead mill type	Outline operating conditions	Ref
Enterobacter cloacae and *Pseudomonas* sp.	Dynomill KDL	Agitator speed: 2000–4000 r.p.m. Glass beads: 0.1–0.45 mm. Bead volume: 80–85% chamber volume. Batch and continuous operation.	20
Saccharomyces cerevisiae	Dynomill KDL and KD-5	45% (wet) (w/v) cell concentration. Laboratory scale and pilot scale disruption. Batch and continuous modes of operation. Key operating parameters investigated.	21
Aspergillus niger, Basidiomyctes	In-house bead mill	About 2 min required for 100% disruption of cells.	22
Arthrobacter sp.	Horizontal grinding chamber (type LME 4, Netzch Feinmahltechnik, Germany)	Cell concentrations of 10–55% (w/v) (wet weight). The effect of power input on disruption investigated.	23

moving beads. Thus, the key operational parameters which influence the process are:

- bead size
- bead volume
- agitator speed
- milling time (or residence time for continuous operations)

The degree of disruption (yield of product) and the rate of intracellular product release can be adjusted by manipulating these variables. There will be a maximum intracellular protein yield (R_M in Equation 1) for each set of conditions, as the more intense the grinding conditions, the more protein will be extracted from membrane and cell wall components. Some typical results are shown in *Figure 2*.

Protocol 4 gives a basic procedure.

Protocol 4

Bead milling Saccharomyces cerevisiae

Equipment and reagents

- Dynomill type KDL (Glen Creston) with 300 ml glass grinding chamber
- 45% suspension of fresh bakers yeast (Mauri Pinnacle Yeasts, Hull) in 10 mM potassium phosphate buffer pH 7.0
- 0.45–0.5 mm glass ballotini beads (Glen Creston)[a]
- Refrigerant at −20°C (50% ethylene glycol) to cool grinding vessel
- Biuret reagent (Merck Ltd.)

Protocol 4 continued

Method

1 Position grinding chamber and ensure a water-tight seal.

2 Add 80% dry bead volume to grinding chamber and select speed of 6000 r.p.m.

3 Begin cooling the grinding vessel and fill the chamber with yeast suspension. Retain a sample on ice for comparative purposes.

4 Run mill in bursts of 15 sec to prevent heating. Maximum protein release occurs within 1–2 min.

5 Centrifuge the resulting suspension at 12 000 g for 30 min at 4 °C to remove cell debris.

6 Assay for protein in supernatant (1:10 dilution) using Biuret reagent.

[a] For bacteria it is beneficial to use a bead size of 0.25–0.3 mm.

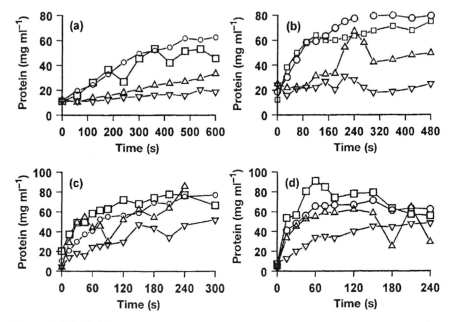

Figure 2 Typical release curves for the disruption of Baker's yeast in a DynoMill. The effects of different bead loadings and agitator speeds are shown. Bead loadings (v/v): (a) 20% (b) 40% (c) 60% (d) 80%. Agitator speeds (r.p.m.): ▽ 2000; $$ = 3000; △ = 4500; ○ = 6000.

2.6 Homogenization

Homogenization describes two sorts of cell disruption processes. The first applies to a hand-held piston/plunger device in which animal cells are easily disrupted. The second type of homogenization refers to disruption in a high pressure homogenizer where the suspension is forced through an orifice.

2.6.1 Homogenization in a piston/plunger device

A procedure is give for the release of enzymes from animal cells, for which this type of homogenizer is suitable.

Protocol 5

Extraction of muscle lactate dehydrogenase from rabbit muscle using a piston homogenizer[a]

Equipment and reagents

- Piston-plunger homogenizer or equivalent
- Rabbit muscle (from recently killed animal)
- Miracloth (Calbiochem)
- Lysis buffer: 30 mM potassium phosphate buffer pH 7.0

Method

1. Mince the tissue if in large pieces.
2. Homogenize the muscle tissue with three volumes of lysis buffer.
3. Stir for 30 min.
4. Centrifuge at 4000 g for 30 min.
5. Filter through Miracloth to remove fat particles.

[a] O'Shannessy et al. (24).

Cell lines have been disrupted by homogenization in a hand-held Wheaton-Dounce homogenizer (size: 1 cm, Wheaton, NJ) with a clearance between the piston and tie of 44–69 μm. However, the method was found to be very inefficient since not even 100 strokes gave the same yield of enzyme as by sonication or by freezing and thawing (25).

2.6.2 Homogenization in a high pressure homogenizer

A group of researchers at UCL (15) were one of the first to use a homogenizer for the disruption of microbes. The homogenizer was originally an instrument for homogenizing milk; but as a result of their work, high-pressure homogenizers more suitable for cell disruption were developed. This type of cell disruption is probably the one of choice for large scale operations. A variety of smaller laboratory scale machines are also available. The homogenizer works by pumping a cell suspension through a narrow orifice at high pressure. This imparts a high velocity on the suspension. A 90° change in direction caused by the valve piston results in the cells crashing into an impact ring through their momentum. Cell breakage occurs from impaction rather than through liquid shear forces. The suspension can be recycled through the machine; each passage through is known as a pass. Usually, several passes are necessary to achieve a satisfactory degree of disruption. The suspension has to be pre-cooled to counter a large rise in temperature associated with passage through the device. Some of the larger machines have an integral cooling system; but it is still advisable to pre-cool the suspension.

i. The effect of operating conditions

The main operating variable is the pressure of homogenization. High pressures are achieved by screwing in the homogenizer valve and thus, leaving only a small annulus for the suspension to flow through. This results in a flow of high-velocity fluid through the annulus which in turn gives a high-impact velocity.

The release of soluble proteins is usually first-order with respect to the number of passes and is modelled by Equation 2, where k was the first-order rate constant. For the disruption of Baker's yeast in a Manton-Gaulin APV homogenizer the value of k is related to the pressure (P) with the form:

$$k = k' P^n \qquad [7]$$

where n is typically about 2.9 and k' is a system constant. The value of n is generally between 1.5 and 2.9 for other cells systems and homogenizers.

Obviously, the greater the number of passes, the greater the degree of cell disruption. However, there is a limit to how small the cell particles can be reduced to by the machine. The location of the enzyme within the cell affects its release by homogenization; and sometimes complete disintegration of the cell will not be required. Some enzymes are released after a relatively small number of passes (4). The homogenizer has not always performed well with mycelial organisms, since the mycelial fragments have been found to block the valve in the homogenizer.

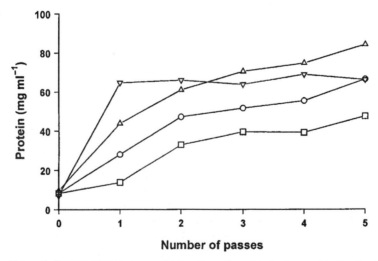

Figure 3 Typical release curves for the disruption of baker's yeast in the Emulsiflex–C5 high pressure homogenizer. It can be seen that increasing the pressure increases the amount and rate of protein release; although the highest pressure reduces the amount of soluble protein released. Homogenizing pressures: □ = 18 k psig; O= 16 k; △=11 k psig; ▽=5 k psig.

Operation of homogenizers will vary according to the manufacturer; but common steps are usually:

(a) Make up the cells in pre-cooled buffer.

(b) Set up the valve setting (and hence pressure) by passing cold water through the homogenizer.

(c) Pump the cell suspension through, making fine adjustments to the valve setting to keep the required operating pressure.

(d) Pass the cells through the homogenizer again, if needed. (Note that it may be necessary to cool the lysate between each passage.)

The setting of the pressure is the crucial operation. High pressures will result in high degrees of breakage on each passage, but will allow only a small flow rate through the machine. Some data from *Protocol 6* are shown in *Figure 3*.

Protocol 6

Laboratory scale homogenization of baker's yeast, Saccharomyces cerevisiae

Equipment and reagents

- Emulsiflex C5 high-pressure homogenizer (Avestin Inc., supplied by Glen Creston Ltd.)
- Biuret reagent

- 45% (w/v) (wet weight) suspension of fresh baker's yeasts (Mauri Pinnacle Yeasts) in 10 mM potassium phosphate buffer pH 7.0

Method

1 Fill sample reservoir with 200 litre of yeast suspension, set the air supply to 100 psig and prime the pump.

2 Adjust the homogenizing valve to give an outlet pressure of 15 k psig.

3 Adjust the flow rate to about 70 cm/ml with valve governing air supply to motor.

4 Recycle homogenate for two to five passes to give maximum disruption. Collect samples in ice between passes.

5 Centrifuge resultant homogenate at 12 000 g at 4°C for 30 min.

6 Assay supernatant for protein (1:10 dilution) with Biuret reagent.

Table 3 shows some operating parameters for different machines and cell types. Of particular importance is the pressure and number of passes.

2.7 Ultrasonication

Ultrasonication is an easy method of disruption for laboratory scale preparations. The vibrating titanium probe creates cavities in the suspension. It is thought that collapse of these cavities results in pressure changes and shear forces which

Table 3 Operating conditions for homogenizers

Cells	Machine	Operating parameters	Ref
Recombinant and non-recombinant *E. coli*	Model M110T, Microfluidics Corp., Newton, MA, USA	Used at 5–50 g/litre dry weight basis. High pressures used: 30–95 MPa. A couple of passes was sufficient. Lysate centrifuged at 10 000 *g* for 30 min.	27
Pseudomonas putida	Manton-Gaulin-APV 15M-8BA Homogenizer	Operating pressure 500 kgf/cm², cell concentration 250 g/litre (wet cells, in phosphate buffer). Up to three passes needed. Cooled to 5 °C. Centrifuged for 1 h at 38 000 r.p.m., 4 °C.	28
Recombinant *E. coli*	Emulsiflex C-30 homogenizer (Avestin Inc., Ottowa)	Cells resuspended to original fermenter concentration. Passed twice through at 8000–10 000 psi.	29
Saccharomyces cerevisiae	Manton-Gaulin-APV homogenizers (15M-8BA and K3)	100–500 kgf/cm². Up to 10 passes, cell concentrations 0.30–0.75 g/ml wet weight.	15
Animal cell lines	1 ml hand-held Wheaton Dunce homogenizer (Wheaton, NJ). Clearance between piston and tube: 44–69 μm).	Clarification at 34 000 *g* for 1 h details not given.	25

cause cell disruption. Ultrasonication is a very vigorous process, and complete solubilization of cells will occur if the process is carried out for a sufficiently long period. This means that cell wall proteins are released in addition to membrane-bound and cytoplasmic proteins. Free radicals may be formed in solution which can result in inactivation of enzymes. The damage caused by these free radicals can be limited by the inclusion of agents such as glutathione or cysteine (30). However, the major problem with ultrasonication is the heat generated during the process. It is usual for the disruption vessels to have a cooling jacket through which coolant is passed (e.g. 10% ethylene glycol solution between −18 and 0 °C). Additionally, if batch ultrasonication is used, it is customary to ultrasonicate in 1 min bursts with a 1 min interval of extra cooling.

Ultrasonicators which have been used for cell disruption have a frequency of around 20 kHz and deliver between 100 and 250 W. Usually it is not possible to adjust the power input. Not all the energy generated by the probe may be available for disruption. Thus, some workers have subtracted the heat energy causing the temperature rise from the acoustic energy. Other workers have assumed that all of the acoustic energy ends up as heat, and that the heat input is equivalent to the acoustic energy input (e.g. ref. 7).

The release curve for intracellular protein is usually first order, with first-order rate constants being in the range of 0.1–2.0 min⁻¹ (Equation 1). The rate constant depends on the power input of the ultrasonicator in an approximate

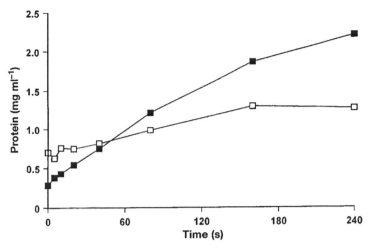

Figure 4 Typical release curves for ultrasonication of baker's yeast in an MSE Soniprep 150. □ = frozen and thawed cells; ■ = fresh cells. It can be seen that there is considerable protein release from the cells just by freezing and thawing. Frozen cells do not break as easily as fresh cells.

linear fashion (31). The rate constant (and thus, the overall rate of the disruption process) will also be dependent on the volume of the sample. The larger the volume of sample, the slower the process. The effectiveness of ultrasonication is usually very dependent on the resonance frequency of the sample. Many machines allow the frequency to be altered until the amplitude is maximum. Note that the high viscosities associated with the disruption of high cell concentrations will reduce the rate constant of disruption.

A protocol for laboratory disruption by ultrasonication is given in *Protocol 7*. Disruption of bacteria and other organisms will be faster or slower according to their cell structure. A typical release curve is shown in *Figure 4*. Some conditions used for ultrasonication are shown in *Table 4*.

Protocol 7

Ultrasonication for the release of protein from *Saccharomyces cerevisiae*

Equipment and reagents

- 20% (w/v) (wet weight) suspension of fresh baker's yeast (Mauri Pinnacle Yeasts) in 50 mM sodium acetate/acetic acid buffer pH 4.7
- Ultrasonicator: Soniprep 150 (MSE)

Method

1 Add 80 ml yeast suspension to a 100 ml, round-bottomed, jacketed vessel with coolant flow (iced water), and immerse probe tip to a depth of 1 cm.

2 Begin sonication, set power to mid-range, and tune to a peak amplitude to18 µm.

3 Sonicate for 1 min intervals allowing 1 min cooling time to elapse between sonication for a total of 40 min.

4 Centrifuge for 30 min at 12 000 g at 4°C to remove cell debris.

5 Assay supernatant for protein by Lowry, with dilution as necessary.

Table 4 Operating conditions for ultrasonication

Cells	Ultrasonicator	Operating conditions	Ref
E. coli	Braun Labsonic 2000 (B. Braun)	Cell concentrations 6–60 g cells/litre (dry weight), frozen at −20°C for 1 week, thawed prior to ultrasonication. Sonication time: 0–20 min. Centrifugation: 12 000 r.p.m. for 45 min at 4°C to remove cell debris.	26
E. coli	Sonic Material, (Vibra Cell)	Ultrasonication combined with chemical treatment. Cell paste suspended in buffer with 0.125% lysozyme. Sonicated for a total of 5 min . Triton N-101 added and stirred for 1 min. Clarification of lysate by centrifugation 20 000 g for 30 min at 37°C.	29
Animal cells (cell lines)	Model W-385, Heat Systems-Ultrasonics, New York (a small 1 ml ultrasonicator)	Frequency used: 20 kHz with 1 sec pulse cycling; 100% release was obtained within 10 sec.	25
Animal cells (cow pancreas pre-ground in centrifugal grinder)	Experimental ultrasonicator	19.5 kHz at 3.3 W/cm^2 (of probe surface) for 5–10 min.	32
Bacillus brevis	Branson S215 sonicator with a 3 mm micro tip (Heat systems, Inc. NY).	Cell suspension in buffer (cells had been frozen −20°C and rethawed). Sonicated in 10 ml batches at 20 kHz for 20 sec intervals cooling to 1–2°C. Centrifuge pellet obtained at 31 000 g for 1 h.	7

2.8 Heat shock

Generally, heat shock is not a suitable method for the release of protein from microbial cells, because thermal denaturation of enzymes can occur, leading to loss of activity and protein precipitation. Rapid heating to 80°C and rapid cooling of *Alcaligenes eutrophus* cells routinely released 10–20% intracellular protein as a soluble form (33); but the key intracellular product (poly-β-hydroxybutyrate) was stable at these temperatures.

2.9 Freezing and thawing

This technique is very simple but works best with cells without a cell wall, such as animal cells. Ice crystals form during freezing and rip the cell membranes and to some extent, cell walls (34). Provided the thawing step is carried out at controlled low temperatures, little damage to enzymes occurs. However, cycles of repeated freezing and thawing, which are often necessary, can denature proteins. The rate of cooling and thawing, as well as the actual temperatures used have been shown to influence the viability of the cells. Animal cells (cell lines) have been disrupted by freezing in an ethanol/dry ice mixture and then thawing in a water-bath at 37 °C. A procedure is shown in *Protocol 8*.

Protocol 8

Freezing and thawing of animal cells[a]

Equipment and reagents

- HeLa cells prepared by trypsinization from monolayer culture, centrifuged, and resuspended in Eagle's medium (DMEM, Gibco)
- Ethanol/dry ice bath
- Water-bath at 37 °C

Method

1 Stir the cells on ice.

2 Freeze cells in ethanol/dry ice bath (3–5 min).

3 Thaw in water-bath at 37 °C (2–3 min).

[a] Shin *et al.* (25).

Storage of cells at −20 °C is commonly done before a specified disruption procedure such as homogenization. The act of freezing and thawing of the cells before passing through the disruption device will often weaken the cells and release a considerable amount of protein. Comparing different data on disruption methods should take into account whether the cells had been frozen beforehand. For instance, just one freezing operation (−20 °C for one week) and subsequent thaw was sufficient to release 21% of the membrane-bound protein, cytochrome b_5, which was released by freezing and thawing plus ultasonication of cells of *E. coli* (26).

2.10 Osmotic shock

Osmotic shock is achieved by incubating cells first in a high osmotic pressure solution and then transferring them to a low osmotic strength buffer. The influx of water bursts the cytoplasmic membrane (35). Gram negative bacteria (*E. coli*) can release their periplasmic protein in this way, but in general osmotic shock does not seem to be a reliable method. Hence, no protocols are given here.

2.11 Lytic enzymes

Lysozyme has been used to hydrolyse β1–4 glycosidic linkages in the peptidoglycan of bacterial cell walls. It is, thus, particularly effective for Gram positive

103

cells, which lyse in minutes. Gram negative cells require a pre-treatment step which allows access of the lysozyme to the cell wall. This is usually achieved by pre-treating the cells with a chelating reagent. Such a procedure is given in *Protocol 9*. Details of the lysis of *Bacillus* are given in *Table 5*.

Table 5 Examples of chemical and enzymic extraction of cells

Cell type (and lysis method)	Details	Ref
Bacillus subtilis with lysozyme and surface active agent	Cells suspended in buffer with sucrose, dithiothreitol, lysozyme (300 µg/ml, EDTA, and Brij 58). Incubated for 1 h on ice. Supernatant obtained after 30 min at 40 000 g.	36
β-Galactosidase release from the yeast: *Kluyveromyces lactis*	2% chloroform, with 10% ethanol. 5–37°C, toluene also tested.	37
Protein and β-lactamase release from *E. coli*	Cell suspension mixed with lysis buffer to give 0.4 M guanidine and 0.5% Triton X-100. Treatment time about 4 h. Clarification by centrifugation.	38

Protocol 9

Enzyme lysis of Pseudomonas putida[a]

Equipment and reagents

- Log phase *Pseudomonas putida* (NCIMB strain 10432) cells, centrifuged 10 000 g for 10 min at 4°C, and stored on ice
- 10 mM potassium phosphate buffer pH 7.0 held at 4°C
- 0.2 M EDTA (tetrasodium salt) held at 4°C
- Lysozyme (Sigma)
- DNase (Sigma) and 0.2 M magnesium chloride
- Lowry protein assay reagents

Method

1 Resuspend 2.5 g wet weight of cells in 10 ml of 10 mM phosphate buffer at pH 7.0 by gentle stirring.

2 Add 50 µl EDTA.

3 Add 2 mg of lysozyme and stir at 30°C for 10 min.

4 Add 300–500 U DNase and 50 µl magnesium chloride. Stir gently for 10 min at 30°C.

5 Centrifuge at 38 000 g for 30 min at 4°C.

6 Assay for protein in supernatant by Lowry's method.

[a] Based on Fish and Lilly (2).

Yeast cell walls can be hydrolysed with snail gut enzymes and β-glucanases from slime moulds or bacteria. If the digestion is performed in a buffer of lower osmotic pressure than the cells, the cells will burst and release their contents. Wiseman and Jones (39) have reported recovering 90% of α-glucosidase and 80%

of lysozyme by snail gut enzymic digestion of yeast. Enzymes from bacteria and slime moulds have been used to release intracellular protein from Brewer's yeast and to a lesser extent, Baker's yeast (40). Asenjo has subsequently carried out further work including modelling the digestion process, and release of recombinant protein (e.g. ref. 41). Whilst lysozyme obtained for egg white is reasonably cheap, the cost of preparation of lytic enzymes obtained from slime moulds is probably more expensive.

The production of lytic enzymes and subsequent lysis of cells occurs naturally in the growth cycle of many microbes during autolysis. It is the basis of the commercial production of hydrolysates such as Marmite (42). However, the process is not really understood enough to be applicable to the release of intracellular enzymes. Cumming *et al.* (18) have found that autolysis is a useful method of cell lysis for *Bacillus amyloliquefaciens* cells. Lysis could be induced by oxygen deprivation, and the process is volume independent.

Lysozyme has been used in combination with mechanical methods as a pretreatment of the cells. See *Table 5* for further details.

2.12 Chemical treatments

The attraction of chemical methods of cell disruption is that the cell will be left substantially intact after release of its contents. This facilitates separation of the cell debris from the supernatant. However, the chemicals used must be compatible with further downstream processes. It will, probably, be necessary to remove the DNA from the lysate, since it is likely to be released in a high molecular weight form, and therefore, greatly increases the viscosity of the lysate.

(a) **Chelating agents** such as EDTA sequest divalent cations. Loss of Mg^{2+} and Ca^{2+} from Gram negative cell walls results in the loss of their permeability barrier. In fact, the inner membrane of *E. coli* remains intact so only periplasmic enzymes are released (35). Chelating agents are not usually used on their own but in combination with another lysis method. Thus, they could be viewed as pre-treatments.

(b) **Chaotropic agents** such as guanidine, ethanol, and urea weaken interaction between hydrophobic molecules. They generally have to be used at high concentrations, which does not make them suitable for large scale extraction processes. Some details are shown in Table 5.

2.13 Detergents

Anionic and non-ionic detergents have been used to permeabilize Gram negative cells. Their site of action appears to be the inner membrane of the cells (35). The action of detergent is often enhanced if the cells are in the exponential phase of growth. Care must be taken to ascertain that the use of a detergent for cell disruption will not interfere with further processes downstream. Non-ionic detergents (e.g. Triton X-100, Brij, Duponal) have all been used to solubilize Gram negative cells.

Protocol 10

Release of membrane-bound enzyme for Nocardia sp. with Triton X-100[a]

Reagents
- Lysis buffer: 0.3 mM phosphate buffer pH 7.5, containing 0.5% Triton X-100
- Culture of *Nocardia*

Method
1 Centrifuge 15 ml of cell culture to obtained a packed cell paste.

2 Add 0.2 g of wet cell paste to 1.8 ml of lysis buffer.

3 Agitate for 45 min at 22°C.

4 Centrifuge the slurry to obtain a clear supernatant containing the enzyme.

[a] Buckland *et al.* (43).

The presence of detergents is often essential to dissolve membrane-bound enzymes (44). Triton X-100 has often been used on its own or in combination with other disruption methods. The 0.2 M guanidine releases periplasmic proteins, treatment with 0.2 M guanidine and 0.5% Triton X-100 releases cytoplasmic protein as well. A method for using Triton X-100 is given in *Protocol 10*, and details of other systems are given in *Table 5*.

Anionic detergents (e.g. SDS at low concentrations, 0.05%) released 24% of intracellular protein (and similar amounts of nucleic acids) from *E. coli* cells (45). Stephenson *et al.* (46) and Rees *et al.* (33) have also reported extensive cell disruption of *E. coli* and *Alcaligenes* cells, respectively with SDS. Examples for SDS lysis of bacterial cells and animal cells are given in *Protocols 11* and *12*.

Addition of specific chemicals to culture broth can induce *E. coli* cells to leak protein into the fermentation broth. Naglak and Wang (38) were able to cause up to 60% of the total protein in *E. coli* to be released within 21 h by exposing exponentially growing cells to 0.4 M guanidine plus 0.5% Triton X-100 at 37°C. However, 75% of the releasable protein could be released with 1 h. Interestingly, although growth is stopped, cells are not lysed.

Protocol 11

SDS lysis of *E. coli* cells[a]

Equipment and reagents
- Suspension of *E. coli* cells in 0.5 M glycerol, 1 mM sodium phosphate at 4°C, made up to an OD (660 nm) of between 10–15
- Ice bath
- Lysis buffer: 2% SDS (v/v) in 0.25 M NaOH with 0.2 M Na$_2$EDTA, stored at 4°C

Method

1   Add 1 ml of lysis buffer to 4.0 ml cell suspension.

2   Mix by inversion once and store on ice.

3   Lysis should occur in about 5 min. The lysate will be very viscous due to the release of DNA, which should remain intact since minimum agitation has been used.

[a] Stephenson *et al.* (46).

Protocol 12

Lysis of animal cells (from cell lines) with SDS[a]

Reagents

- Chloroform
- Cell suspension
- 0.1% SDS

Method

1   Add 15 μl of chloroform and 15 μl of 0.1% SDS to 1 ml of cell suspension.

2   Agitate the mixture for 10 sec.

3   Centrifuge at 12 000 g for 15 min at 4°C.

4   Assay of protein can be carried out but there is interference from SDS. SDS of the same concentration needs to be included in blanks and standards.

[a] Shin *et al.* (25).

A combination of chemical agents with mechanical methods has been successfully used to increase the yield or rate of the disruption process. For example, alkali or detergent pre-treatment of *Alcaligenes eutrophus* cells enhanced release of polyhydroxybutyate (47).

2.14 Solvents

A number of solvents have been used for the release of intracellular components from bacteria and yeast. Provided that the solvent does not inactivate the enzyme product, it is a simple method. Toluene solubilizes the inner membrane of *E. coli* to various degrees according to its final concentration. Thus, low concentrations (< 5%) cause small molecular weight materials to be released from cells. Higher concentrations can release 20–30% of the intracellular protein. Toluene renders Gram positive cells and yeasts permeable to other agents without releasing large amounts of intracellular material. Ether, isoamyl alcohol, and chloroform have also been used for *E. coli* protein release. *Table 5* gives some details, but consult ref. 38 for a detailed discussion.

3 Conclusions: choice of methods

(a) The mechanical methods of cell disruption have the widest application for laboratory and pilot scale disruption.

(b) Homogenization has proved an effective large scale process.

(c) Chemical methods are, generally, cell/product specific and thus, not applicable to all systems.

References

1. Gonzales, C., Lagos, R., and Montasterio, O. (1996). *Micobios*, **85**, 205.
2. Fish, N. M. and Lilly, M. D. (1984). *Bio/Technology*, **July**, 623.
3. Middelberg, A. J. (1995). *Biotechnol. Adv.*, **3**, 491.
4. Follows, M., Hetherington, P. J., Dunnill, P., and Lilly, M. D. (1972). *Biotechnol. Bioeng.*, **13**, 549.
5. Chi, W. K., Ku, C. H., Chang C. C., and Tsa, J. N. (1997). *Ann. N. Y. Acad. Sci.*, **721**, 365.
6. van Gaver, D. and Huyghebaert, A. (1990). *Enzyme Microb. Technol.*, **13**, 665.
7. Augenstein, D. C., Thrasher, K., Sisnskey, A. J., and Wang, D. I. C. (1974). *Biotechnol. Bioeng.*, **15**, 1433.
8. Jazwinski, S. M. (1990). In *Methods in enzymology* (ed. M. P. Deutscher), Vol. 182, p. 154. Academic Press, London.
9. Hedenskog, G. and Ebbinghaus, L. (1972). *Biotechnol. Bioeng.*, **14**, 447.
10. Mosqueira, F. G., Higgins, J. J., Dunnill, P., and Lilly, M. D. (1981). *Biotechnol. Bioeng.*, **23**, 335.
11. Clarkson, A. I., Lefevre, P., and Titchener-Hooker, N. J. (1993). *Biotechnol. Prog.*, **9**, 462.
12. Herbert, D., Phipps, P. J., and Strange, R. E. (1971). In *Methods in microbiology 5B* (eds. J. R. Norris and D. W. Ribbons), pp. 209–344. Academic Press, London, New York.
13. Davies, R. (1959). *Biochim. Biophys. Acta*, **33**, 481.
14. Kuboi, R., Umakoshi, H., Tfafgi, N., and Komosawa, I. (1995). *J. Ferm. Bioeng.*, **79**, 335.
15. Hetherington, P. J., Follows, M., Dunnil, P., and Lilly, M. D. (1971). *Trans. Inst. Chem. Eng.*, **49**, 142.
16. Dunnill, P. (1982). *Chem. Ind.*, **1982**(2), 877.
17. Foster, D. (1995). *Curr. Opin. Biotechnol.*, **6**, 523.
18. Cumming, R. H., Tuffnel, J., and Street, G. (1985). *Biotechnol. Bioeng.*, **27**, 887.
19. Gegenheimer, P. (1990). In *Methods in enzymology* (ed. M. P. Deutscher), Vol. 182, p. 174. Academic Press, London.
20. Woodrow, J. R. and Quirk, A. V. (1982). *Enzyme Microbiol. Technol.*, **4**, 385.
21. Limon-Lason, J., Hoare, M., Orsbo, C. B., Doyle, D. J., and Dunnill, P. (1979). *Biotechnol. Bioeng.*, **21**, 745.
22. Rehacek, J., Beran, K., and Bicik, V. (1969). *Appl. Microbiol.*, **17**, 462.
23. Bunge, F., Pietzsch, M., Müller, R., and Slydatk, C. (1992). *Chem. Eng. Sci.*, **47**, 225.
24. O'Shannessy, K., Scoble, J., and Scopes, R. K. (1996). *Bioseparation*, **6**, 77.
25. Shin, J. H., Lee, G. M., and Kim, J. K. (1994). *Biotechnol. Tech.*, **8**, 425.
26. Santos, J. A. L., Belo, I., Mota, M., and Cabral, J. M. S. (1996). *Bioseparation*, **6**, 81.
27. Sauer, T., Robinson, C. W., and Glick, B. R. (1989). *Biotechnol. Bioeng.*, **33**, 1330.
28. Fish, N. M., Harbron, S., Allenby, D. J., and Lilly, M. D. (1983). *European J. Appl. Microbiol. Biotechnol.*, **17**, 57.
29. Caldwell, S. R., Varghese, J., and Puri N. K. (1996). *Bioseparation*, **6**, 115.

30. Garcia, F. A. P. (1993). In *Recovery processes for biological materials* (ed. J. F. Kennedy and J. M. S. Cabral), p. 47. J. Wiley, New York.

31. Doulah, M. (1977). *Biotechnol. Bioeng.*, **19**, 649.

32. Zayas, J. (1985). *Biotechnol. Bioeng.*, **27**, 1223.

33. Rees, P., Watson, J. S., Cumming, R. H., Liddell, J. M., and Turner, P. D. (1996). *Bioseparation*, **6**, 125.

34. Scopes, R. K. (1982). In *Protein purification: principles and practice*, p. 21. Springer–Verlag, New York.

35. Naglak, T. J., Hettwar, D. J., and Wang, H. Y. (1990). In *Separation processes in biotechnology* (ed. A. Asenjo). Marcel Dekker, Inc., NY.

36. Cull, M. and McHenry, C. S. (1990). In *Methods in enzymology* (ed. M. P. Deutscher), Vol. 182, p. 148. Academic Press, London.

37. Flores, M. V., Voget, C. E., and Ertola, R. J. J. (1994). *Enzyme Microb. Techn.*, **16**, 340.

38. Naglak, T. J. and Wang, H. Y. (1992). *Biotechnol. Bioeng.*, **39**, 732.

39. Wiseman, A. and Jones, P. R. (1971). *J. Appl. Chem. Biotechnol.*, **2**, 26.

40. Asenjo, J. A. and Dunnill, P. (1981). *Biotechnol. Bioeng.*, **23**, 1045.

41. Asenjo, J. A., Andrews, B. A., and Pitts, J. M. (1988). *Ann. N.Y. Acad. Sci.*, **542**, 140.

42. Reed, G. and Nagodawithana, T. W. (1991). In *Yeast technology*, Chap. 8, p. 370. Van Nostrand Reinhold, NY.

43. Buckland, B. C., Lilly, M. D., and Dunnill, P. (1974). *Biotechnol. Bioeng.*, **18**, 601.

44. Neugebauer, J. M. (1990). In *Methods in enzymology* (ed. M. P. Deutscher), Vol. 182, p. 182. Academic Press, London.

45. Woodringh, C. L. (1970). *Biochim. Biophys. Acta*, **224**, 288.

46. Stephenson, D., Norman, F., and Cumming, R. H. (1993). *Bioseparation*, **3**, 285.

47. Harrison, S. T. L., Dennis, J. S., and Chase, H. A. (1991). *Bioseparation*, **2**, 95.

Chapter 6
Concentration of the extract

E. L. V. Harris

Bioprocessing Consultant, 20 Codmore Crescent, Chesham, Buckinghamshire
HP5 3LZ, UK.

1 Introduction

A concentration step is frequently required after a clarified solution of the protein has been obtained, in order to aid subsequent purification steps. This is particularly important when the protein is obtained in culture medium from cells (e.g. bacteria or tissue culture cells). Concentration of the protein solution results in a decreased volume, as well as a higher protein concentration. Clearly a smaller volume of solution is easier to handle in subsequent steps, such as precipitation or loading onto a chromatography column. Higher protein concentration minimize protein losses by non-specific adsorption to container walls or column matrices. In addition many subsequent purification steps require a minimum protein concentration to be effective, for example, precipitation is more efficient at concentrations above 100 µg/ml, whilst for adsorption chromatography (e.g. ion exchange or affinity) the concentration of protein must be greater than the dissociation constant.

Concentration is achieved by removal of water and other small molecules:

(a) By addition of a dry matrix polymer with pores that are too small to allow entry of the large protein molecules (Section 2).

(b) By removal of the small molecules through a semi-permeable membrane which will not allow the large molecules through (i.e. ultrafiltration, Section 3).

(c) By removal of water *in vacuo* (i.e. lyophilization, Section 4).

Precipitation can also be used to concentrate proteins if the pellet is redissolved in a smaller volume, and in addition often results in some degree of purification of the protein of interest. However, as mentioned above precipitation is more effective if the total protein concentration is above 100 µg/ml (see Section 6). Two-phase aqueous extraction can also be used to concentrate the protein, with an associated degree of purification (see Section 7).

2 Addition of a dry matrix polymer

This is one of the simplest and quickest methods of concentrating solutions of proteins, requiring minimal apparatus. A dry matrix polymer, such as Sephadex, is added to the protein solution and allowed to absorb the water and other small molecules; the pores within the matrix are too small to allow the protein to be absorbed. When the matrix has swollen to its full extent the remaining protein solution is removed after the matrix has been settled by gravity, filtration, or centrifugation. A method for this is given in *Protocol 1*. The degree of concentration obtained by this method is low compared to some of the other methods. Another major disadvantage of this method is that not all the protein solution can be recovered, since some will be trapped in the matrix bed, between the matrix particles, resulting in a low yield (at best 80–90%, depending on the volume of the matrix bed). Thus, this method is not widely used unless concentration is of more importance than the yield of the protein.

Protocol 1

Concentration using dry Sephadex

Equipment and reagents
- Centrifuge
- Whatman No. 1 filter paper
- Sephadex G-25

Method

1 Add dry Sephadex G-25 to the solution to be concentrated at ratio of 20 g Sephadex per 100 ml solution.

2 Allow the Sephadex to swell for ∼ 15 min.

3 Remove the supernatant by either:
  (a) Centrifuging at 2000 g for 10 min.
  (b) Filtration through Whatman No. 1 filter paper. Typically the volume of supernatant will be ∼ 30% of the original volume and contain ∼ 80% of the protein.

Matrices have been developed which have a temperature-dependent hydrated volume (1). These may prove useful in large scale applications such as the biotechnology industry. The dry matrix is added at a carefully controlled temperature and the bulk of the protein solution removed after the water has been absorbed; the temperature is raised by as little as 1 °C resulting in rapid shrinking of the volume occupied by the matrix of up to tenfold. The released solution can then be removed and the matrix more rapidly dried for subsequent use.

3 Ultrafiltration

In ultrafiltration, water and other small molecules are driven through a semipermeable membrane by a transmembrane force such as centrifugation or high pressure. For ultrafiltration, the membrane pores range in diameter from 1–20 nm; the diameter is chosen such that the protein of interest is too large to pass through. Pore sizes of microfiltration membranes range from 0.1–10 μm diameter and allow proteins and other macromolecules to pass through, whilst retaining larger particles such as cells. Rather than quote the pore size of the ultrafiltration membranes it is more common to quote a nominal molecular weight cut-off (NMWC) for the membrane. The NMWC is defined as the minimum molecular weight globular molecule which will not pass through the membrane. It is important to remember that the shape of the molecule will affect whether it can pass through the pores, thus whilst a globular protein of 100 000 mol. wt. will not pass through a 100 000 NMWC membrane, linear molecules of 1 000 000 mol. wt. have been shown to pass through under certain conditions (2). The pore size are not uniform and will show a normal distribution around the mean pore size (*Figure 1*). The range of this distribution will vary with the method of manufacture of the membrane, and therefore also between manufacturers. A tenfold change in pore size will result in a 100-fold change in the NMWC. Thus, the NMWC of the membrane used should usually be significantly less than the molecular weight of the protein of interest (usually ≥ 20% less). If too small a NMWC is chosen the flow rates through the membrane will be reduced, resulting in longer times or a requirement for a higher driving force.

Concentration by ultrafiltration offers several advantages over alternative methods.

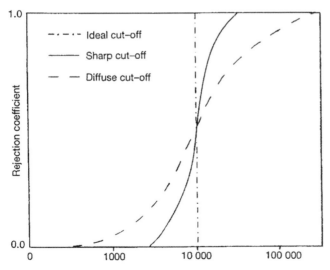

Figure 1 Rejection characteristics of a membrane with a NMWC of 10 000. Ideally all molecules of ≥ 10 000 molecular weight will not pass through the membrane whilst those of < 10 000 will. However, due to the distribution of pore sizes in membranes there is also a distribution in molecular weight of the molecules able to pass through the membrane. Thus, for a particular membrane the filtrate will contain a certain percentage of molecules with molecular weight less than the NMWC and a similar percentage with a higher molecular weight.

(a) Precipitation followed by centrifugation requires a minimum concentration of 100 μg/ml and often results in poor recoveries due to the phase change. In addition the volumes which can be handled easily are limited (particularly if continuous flow centrifuges are unavailable), and aerosols are hard to contain.

(b) Concentration by dialysis requires longer processing times and volumes are severely limited by ease of handling.

(c) Freeze-drying requires longer processing times and can result in poor re-coveries due to the phase change. In addition to concentrating the protein, salts are also concentrated.

Equations which fully describe the observed behaviour of microfiltration and ultrafiltration processes have not yet been devised. The reader is referred to refs 3–5 for discussion of the mathematical models; the qualitative principles are discussed below. The flux rate across the membrane (i.e. the volume of filtrate per unit area per unit of time) is the main factor requiring optimization, since a

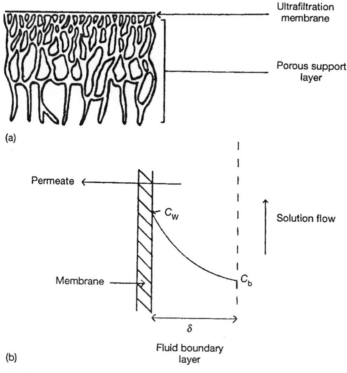

(a)

(b)

Permeate

Membrane

C_w

C_b

Solution flow

δ

Ultrafiltration membrane

Porous support layer

Fluid boundary layer

Figure 2 (a) Asymmetric ultrafiltration membranes consist of a thin upper ultrafiltration layer and a deep porous support layer. (b) Concentration polarization is caused by a build-up of solute molecules at the surface of the membrane. A solution of concentration C_b flows across the membrane. Water and small solute molecules are forced across the fluid boundary layer (thickness δ) and pass through the membrane. Larger solute molecules, which cannot pass through the membrane, concentrate at the membrane surface. A concentration gradient of these solute molecules therefore occurs within the fluid boundary layer. This concentration gradient is termed concentration polarization.

higher flux rate results in shorter processing time. Flux rate is directly proportional to the transmembrane pressure (up to a certain limit) and indirectly proportional to the resistance against passage of molecules across the membrane. This resistance is the sum of three factors: the membrane resistance, and the resistances caused by concentration polarization and fouling (see below).

The resistance of the membrane is minimized by the following:

(a) Increased pore size, hence the maximum pore size which does not let the protein of interest pass through should be chosen.

(b) Increased pore density, which will vary from manufacturer to manufacturer, and often batch to batch.

(c) Minimal thickness of the membrane, hence most ultrafiltration membranes are asymmetric (or anisotropic). These ultrafiltration membranes consist of a thin upper layer (~ 0.5 μm thick) of small defined pore size (i.e. the ultrafiltration membrane itself) with a deep, large pore, spongy lower layer (i.e. a support layer) about 150 μm thick (*Figure 2a*).

(d) Maximum wettability of the membrane.

(e) Minimum viscosity of the solution. Viscosity is indirectly proportional to temperature and directly proportional to the concentration of solutes.

Concentration polarization is caused by a build-up of molecules at the surface of the membrane (*Figure 2b*) (6–8). A gel-like layer forms and acts as a second ultrafiltration layer decreasing the flux rate and preventing passage of some molecules which would normally pass through the membrane. The build-up of this layer is minimized by allowing the molecules to diffuse back into the solution by agitating the solution close to the membrane. This is achieved either by stirring as in stirred cells or by tangential or cross-flow (*Figure 3*).

Fouling of the membrane is caused by particles, or macromolecules becoming adsorbed to the membrane, or physically embedded in the pores. Once a membrane is fouled, cleaning will not normally return the flux rate to its initial value, whilst in contrast cleaning will remove the layer caused by concentration polarization. A simple way to distinguish between whether the decrease in flux rate is caused by concentration polarization or by fouling is to observe the effect

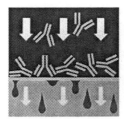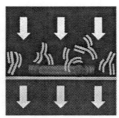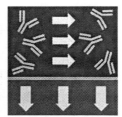

Figure 3 With dead-end filtration a gel-like layer rapidly builds up at the membrane surface (a) and severely reduces flux rates. The build-up of this layer is minimized by stirring as in stirred cells (b) or by tangential or cross-flow (c). (Courtesy of Millipore.)

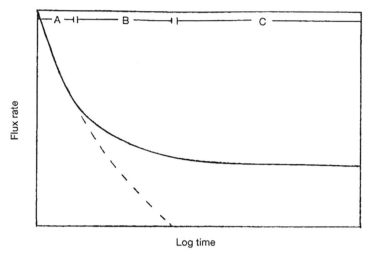

Figure 4 Decay of flux rate across the membrane with time. (A) The initial flux rate decreases as concentration polarization occurs. (B) As solute concentration increases the viscosity increases and causes a decrease in flux rate. (C) Finally a steady state flux is achieved. When fouling occurs (—) this steady state flux is never achieved. Typically phases A–B occur within 5–10 min.

of increasing the tangential flow on the flux rate. If increasing the tangential flow increases the flux rate then concentration polarization was the problem: if increasing the tangential flow decreases the flux rate then it was fouling. Observation of the decay in flux rate with time will also distinguish between concentration polarization and fouling. The flux rates through a membrane will follow a sigmoidal decrease (*Figure 4*). With a new or clean ultrafiltration membrane the initial flux is relatively high and will rapidly decrease as concentration polarization occurs. An equilibrium will be reached and maintained until the concentration, and hence the viscosity increases; at equilibrium the cleaning caused by tangential flow equals the rate of concentration polarization. If fouling occurs, the flux rate will continue to decrease with time and an equilibrium will not be achieved. Fouling can be minimized by removing particulate matter prior to ultrafiltration and by choosing conditions where the macromolecules in the solution do not interact with the membrane (thus pH, ionic strength, and membrane type are important things to consider).

3.1 Equipment

3.1.1 Small scale concentration

Although not strictly defined as ultrafiltration, the principles of concentration by dialysis are similar and are therefore covered here. The protein solution to be concentrated is placed in a bag of dialysis tubing (see Section 5.1), which is placed in a solution or powder that draws water through the dialysis membrane. Solutions of polyethylene glycol (PEG) (mol. wt. $\geq 20\,000$) at a concentration of 20% (w/v) are frequently used.

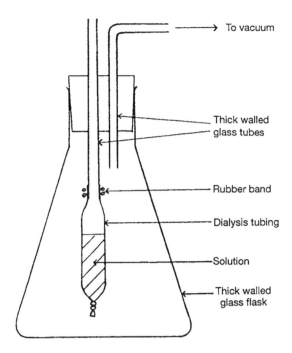

To vacuum

Thick walled
glass tubes

Rubber band

Dialysis tubing

Solution

Thick walled
glass flask

Figure 5 Apparatus for vacuum dialysis. NB: ensure that the tubes and flask used are capable of withstanding a vacuum.

Alternatively, a dry matrix polymer such as Sephadex can be used. The former method has the advantage that the concentration can be left unattended without fear of complete removal of the water, which could result in loss of the protein by irreversible absorption to the dialysis membrane. However, PEG may contain small molecular weight impurities that can inactivate enzymes. Water can also be 'sucked' through the dialysis membrane by applying a vacuum to the outside of the dialysis bag whilst the inside of the bag is maintained at atmospheric pressure. For this technique the apparatus shown in *Figure 5* is used. These methods of concentration are very suitable for small volumes (≤ 50 ml), practical details are given in *Protocol 2*.

Several types of apparatus are available commercially for ultrafiltration on a small scale, combining use of centrifugal force with an absorber pad or ultrafiltration through a semi-permeable membrane (e.g. Centricon, Millipore and Vectaspin, Whatman, see *Figure 6*). The semi-permeable membrane is available in a range of pore sizes, with molecular weight cut-offs ranging from 3–100 kDa. Care should be taken to note the chemical compatibility indicated by the manufacturer, particularly with regard to pH and organic solvents. Both of these types of apparatus are convenient for small volumes (≤ 5 ml) of multiple samples and are therefore more often used for concentration, or removal of small molecular weight contaminants prior to analysis of proteins, rather than for purification.

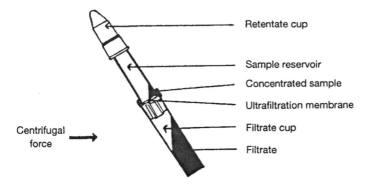

Retentate cup

Sample reservoir

Concentrated sample

Ultrafiltration membrane

Centrifugal force →

Filtrate cup

Filtrate

Figure 6 Centricon shown after concentration. To recover the retentate, the sample reservoir and retentate cup can be removed from the filtrate cup, inverted, and centrifuged, thus the retentate is obtained in the retentate cup.

Protocol 2

Concentration by dialysis against PEG or dry Sephadex

Equipment and reagents

- Dialysis tubing
- Dialysis bag
- Magnetic follower and stirrer
- PEG 2000 solution
- Sephadex G-25

Method

1 Treat the dialysis tubing as described in *Protocol 5*.

2 Place the solution to be concentrated in a dialysis bag sealed at one end with a knot or plastic clip.

3 Expel the air from the dialysis bag and seal the remaining end of the bag. Place the bag in a container of either:

   (a) 20% (w/v) PEG 2000 solution, 5–10 × the volume of the solution to be concentrated.

   (b) Dry Sephadex G-25, ~ 20 g per 100 ml of solution.

4 Stir the PEG solution using a magnetic follower and stirrer, or occasionally remove the swelling Sephadex from the dialysis bag and agitate.

5 Incubate until the desired concentration is reached. With the Sephadex it is important to ensure that the water is not completely removed, and therefore the volume in the dialysis bag should be checked about every 30 min. With the PEG solution the concentration may be left unattended overnight.

For larger scale applications high flow rates are desirable to minimize process time. These are achieved (i) by the use of asymmetric (or anisotropic) membranes; and (ii) by agitating the solution close to the membrane by stirring or tangential flow to minimize membrane fouling and concentration polarization.

3.1.2 Stirred cells

Stirred cell are available to cover the range 1–400 ml (e.g. Millipore) (*Figure 7*). Several types of membranes are available with NMWCs from 500–1 000 000. YM membranes (Millipore) are particularly useful for concentrating dilute solutions of proteins due to their low non-specific binding properties. To minimize losses the smallest surface area of membrane should be used which will still allow a reasonable flow rate. It is advisable not to allow the solution to dry out completely, since this can cause irreversible loss of the protein onto the membrane surface. After use, membranes can be cleaned with dilute NaOH, 1–2 M NaCl, or a dilute surfactant (check manufacturer's recommendations), washed with water, and stored in 10% ethanol at 4°C. A protocol for using stirred cells is given in *Protocol 3*.

3.1.3 Tangential or cross-flow systems

Stirred cells are easy and convenient to use at the laboratory scale, but cannot be used at process scale because of the large surface areas of membrane required to achieve suitable flow rates. There are several types of ultrafiltration systems and membranes suitable for large scale use. These fall into four categories.

(a) **Flat plate** (*Figure 8a*). Flat sheets of membrane are stacked between stainless steel or acrylic plates. The membranes may be individual or stacked in cassettes. The solution is pumped tangentially across the membrane or membrane stacks and is recycled (retentate), whilst filtrate passes through the

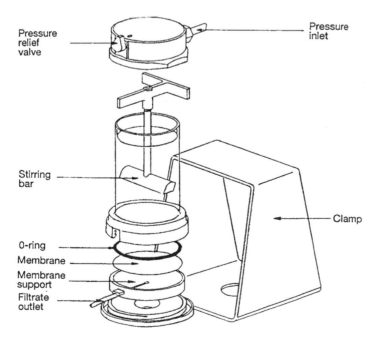

Figure 7 Stirred cell.

membrane and is channelled along a separate flow path to the collection vessel. Scale-up is easily achieved by increasing the number of membranes in a stack, or by connecting stacks together, thus increasing the membrane surface area. These systems can be used to process volumes of 200 ml or more.

(b) **Spirals** (*Figure 8b*). Several flat sheets of membrane are sandwiched between spacer screen and then the whole stack is wound spirally around a hollow, perforated cylinder. The solution is pumped parallel to the long axis of the cylinder, the filtrate (permeate) passes through into a collection channel which is connected to the central hollow cylinder. The retentate exits from the spiral cartridge and is recirculated until the desired concentration has been achieved. Spiral membrane systems can be used for volumes of a few litres upwards.

(c) **Hollow fibres** (*Figure 8c*). The membrane is produced as a self-supporting hollow fibre with an internal diameter of 0.5–3.0 mm. The fibres are assembled in bundles in a cylindrical cartridge. The ultrafiltration membrane is usually on the inside of the fibres, in which case the solution is pumped through the fibres and the filtrate (permeate) passes out into the cylindrical cartridge. Hollow fibre systems are available to cover a wide range of volumes from 25 ml upwards by using different cartridge sizes and linking several cartridges together.

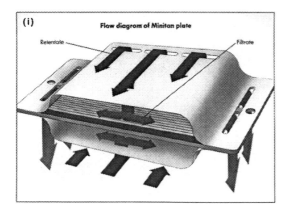

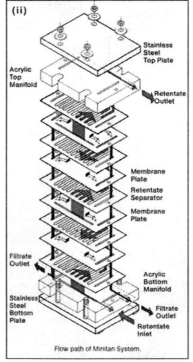

Figure 8a Flat plate ultrafiltration system showing an individual plate (i) and the complete Minitan system (ii); (courtesy of Millipore).

(d) **Tubes**. These are similar to hollow fibres but have much larger diameters (typically 2–3 cm) and consequently much larger internal volume to surface area ratios. Because of these larger ratios the flux through the membranes is lower than with hollow fibres, and therefore tube systems are not as widely used.

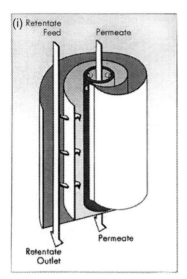

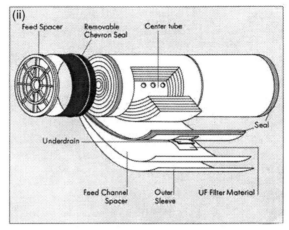

Figure 8b Spiral ultrafiltration system showing a flow schematic (i) and the complete cartridge (ii); (courtesy of Millipore).

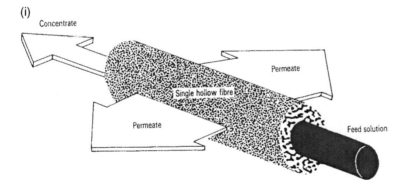

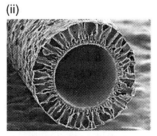

Figure 8c Hollow fibre ultrafiltration system showing a flow schematic (i) and an electron micrograph of an individual fibre (ii) (reproduced with permission from Romicon).

Protocol 3

Ultrafiltration using stirred cells

Equipment

- Ultrafiltration membrane
- Magnetic stirrer
- Centrifuge

Method

1   Pre-treat the ultrafiltration membrane according to the manufacturer's instructions to remove preservatives. Typically soak the membrane in three changes of water for ~ 1 h total.

2   Assemble the stirred cell according to the manufacturer's instructions. Ensure the membrane is placed the correct way up (shiny surface uppermost).

3   Pre-filter or centrifuge the solution to remove particulate matter and gently pour into the stirred cell. Attach the cap assembly and place in the retaining stand.

4   Ensure the pressure relief cap is closed and attach to a regulated pressure source, typically a nitrogen cylinder. Place the assembly on a magnetic stirrer.

5   Apply the minimum pressure required to give an acceptable flow rate; higher pressures will lead to increased concentration polarization and fouling, and therefore a reduced flux rate. Do not exceed the maximum pressure recommended by the manufacturer. Place the filtrate tubing into a collection vessel.

6   Turn on the magnetic stirrer and adjust the stirring rate so that the vortex is no more than one-third the depth of the solution. Excess stirring will denature the protein by shear forces and foaming; insufficient stirring will cause increased concentration polarization.

7   When the desired concentration has been achieved turn off the pressure and open the pressure relief valve. Continue stirring for 5 min to resuspend the polarized layer.

8   Remove the cap assembly and gently pour out the concentrate.

Several factors influence the choice of membrane configuration: hold-up volumes; flux rates; sensitivity to fouling; ease of cleaning; and ease of scale-up (Table 1). Systems based on the flat sheet configuration have minimal hold-up volumes and are therefore particularly useful for lower volume work, and where minimal losses are required. The flux rate is proportional to the ratio of surface area to volume, thus tube systems give poorer flux rates and are therefore less economical due to the longer processing times. Tube systems are, however, the least sensitive to fouling by particulate matter in the retentate stream and are therefore particularly useful for concentrating turbid solutions, or for clarification by microfiltration. Spirally wound membranes are not recommended for use with sample streams containing particulate matter. Hollow

Table 1 Properties of different ultrafiltration membrane configurations

	Pre-filtration required	Hold-up volume	Power consumption	Compactness	Other properties
Flat plate	Yes	Low	Medium	Medium	Withstands high pressure differentials; suitable for highly viscous solutions; steam sterilizable *in situ*
Spiral cartridge	Yes	Low	Low	High	Not suitable for particulate or viscous solutions
Hollow fibre	Yes	Low	Low	High	Can be back-flushed to ease cleaning; will not withstand high pressure differentials
Tubular	No	High	High	Low	Poor flux rates; require high pumping rates; suitable for particulate solutions

fibre membranes are the most effectively and easily cleaned, since they can be back-flushed to remove fouling material. The most widely available systems are those based on hollow fibre membranes or flat sheets, followed by spirally wound membranes.

With all these membrane configurations a pump is required to recirculate the retentate through the system until the desired concentration has been reached. Suitable pumps are supplied by the manufacturers of the ultrafiltration system. Different types of pump are available: peristaltic rotary piston, rotary lobe, diaphragm, progressing cavity, piston, and centrifugal. Choice will be influenced by several factors, such as cost, pressure required, hold-up volume, shear sensitivity of protein, period of continuous use, and requirement for steam sterilization. Peristaltic pumps are relatively cheap, have minimal hold-up volumes and give minimal shear. However, they are only suitable for use at less than 2 bar and, unless heavy duty, can only be used for continuous periods of less than 2 h. The hoses must be carefully watched as they are prone to failure. Centrifugal pumps can be used at less than 5 bar, but are the most likely pumps to cause shear-induced denaturation and have high hold-up volumes. Diaphragm pumps are less likely to cause shear-induced denaturation, but can be hard to scale-up. For pressures of more than 10 bar piston, diaphragm or multistage centrifugal pumps are suitable. Most manufacturers tend to prefer positive displacement rotary lobe pumps running at well below their maximum rated speeds.

3.1.4 Membrane composition

Traditionally all ultrafiltration membranes have been made from cellulosic materials (e.g. cellulose acetate). However, these materials have limited chemical and thermal stability, and therefore cannot be used for all applications. Several plastic membrane types have become available, such as polycarbonate, polyamides, polysulfone, polyvinyl chloride, and acrylonitrile polymers. These mem-

123

branes are compatible with a wider range of chemical conditions than the cellulosic membranes, and can be used over the pH range 1–13 and with a variety of organic solvents. They are also more stable to heat and can be used at up to 80 °C. A few membranes can be autoclaved, but this often results in an increase in the NMWC. Many manufacturers give little information about the composition of their non-cellulosic membranes, however it appears that poly-sulfone is currently the most commonly used material. There are a few inorganic-based membranes available which have a support layer of carbon with an ultrafiltration layer of zirconium oxide. These inorganic membranes have extremely good chemical and thermal resistance, and can be operated at 100 °C and sterilized by autoclaving.

3.2 Operation

3.2.1 Optimization

One of the key parameters requiring optimization is the flux rate, particularly for large scale applications. In addition denaturation and loss of the protein must be minimized and, if appropriate, selectivity should be maximized to achieve purification in addition to concentration.

Although the choice of factors to consider during optimization (see below) can be minimized by careful theoretical consideration, final optimization can only be determined by experimental studies. An ideal optimization would follow the protocol given in *Protocol 4*.

Important factors to consider in optimizing a process were briefly discussed in Section 3 and are discussed in more detail here.

(a) **Transmembrane pressure**. Maximum transmembrane pressure will give maximum flux rate, but above a certain limit concentration polarization will limit the flux rate and fouling may be enhanced. Pumping costs will also increase.

(b) **Tangential flow**. Maximum tangential flow will minimize concentration polarization and fouling but may cause denaturation of the protein by shear-stress (10, 11) and will increase pumping costs.

(c) **Viscosity of the solution**. High viscosity causes decreased flux rate (12). Viscosity can be minimized by increasing the temperature, however, since most proteins are denatured by heat the practical limit is usually ∼ 40 °C. DNA is often a cause of high viscosity, therefore prior treatment with DNase will often increase the flux rate. As the concentration of the protein increases the viscosity will increase and therefore decrease the flux rate.

(d) **Composition of the solution**. Particulate matter should usually be removed prior to ultrafiltration. This is best done by centrifugation, flocculation, coarse filtration, or microfiltration. Antifoam, often included in fermentation media, increases concentration polarization and fouling (2). The magnitude of the effect varies from type to type; the silicone emulsions have the greatest effect. Thus the antifoam should be chosen with care and its concentration minimized.

pH and ionic strength can also have a profound effect on the flux rate. The buffer conditions (i.e. pH and ionic strength) should be chosen to minimize precipitation of any of the sample components in order to minimize membrane fouling. This is especially important for the protein of interest, since precipitation onto or into the membrane will cause irreversible losses. The pH should usually not be close or equal to the pI of the protein (13), since the flux rate will be minimal. Ionic strength may affect the aggregation of a protein and may therefore also affect the observed rejection characteristics (9).

(e) **Choice of membrane and system**. Flux rate increases with size and number of pores. In order to choose an appropriate membrane the flux rate should be measured experimentally for each application. The experimentally observed NMWC may differ significantly from the manufacturer's claimed NMWC and both size distribution and numbers of pores may vary from batch to batch (9). The susceptibility to fouling also varies with membrane type and manufacturer (9). The degree of non-specific adsorption of the protein to the membrane varies with membrane type, from protein to protein, and with buffer conditions and protein concentration. As an example, polysulfone membranes have been reported to show higher adsorption of bovine serum albumin (BSA) than cellulose acetate membranes (14). Cleaning can be done *in situ* with many systems, or after dismantling. The manufacturer's recommendations for cleaning should be followed to avoid damage of the membrane or system. Some recommended cleaning solutions are 5% NaCl, 1 M HCl, or 1 M NaOH; some manufacturers supply proprietary cleaning agents. Use of elevated temperatures may enhance the cleaning procedure. Hollow fibre configurations can be back-flushed to maximize the efficiency of cleaning; particularly useful if the membrane has been fouled. After cleaning, the system and/or membrane should be thoroughly rinsed with water to remove the cleaning agent. The efficiency of cleaning can be assessed by measuring the flux rate, which should be restored to 95% or more of the initial flux rate of the new membrane (2, 15).

Protocol 4

Optimization of ultrafiltration

1   Assemble the system according to the manufacturer's instructions. Ideally use a new membrane. Check for leaks.

2   Pump clean water through the system at an appropriate transmembrane pressure and tangential flow rate. Measure the flux rate whilst recycling the filtrate, either with an in-line flow meter on the filtrate line or by measuring the volume of filtrate collected over a given period of time (e.g. 1 min). It is important to use water purified by reverse osmosis, since fouling may occur with inferior quality water (9). Continue measuring flux rate for at least 15 min and plot against time. Flux rate should not decrease significantly with time.

Protocol 4 continued

3   Check membrane integrity by either of the following methods. Use a solution (~ 1%) of pure protein which will not pass through the pores. Pump this solution through the system and check the filtrate for absence of protein by a standard protein assay. Before continuing to step 4 clean the membrane as described in steps 8 and 9. Alternatively, attach a supply of nitrogen to the system and pressurize to ~ 10 psi. Monitor the decrease in pressure with time; a slow decrease is expected due to diffusion. If a leak is present in the membrane or system the pressure will not be maintained. Alternatively measure the volume of water displaced into the filtrate stream; typically this will be 1–2 ml/min/ft^2 membrane.

4   Pump the solution of interest through the system allowing the permeate to recycle into the retentate. Initially use the highest tangential flow achievable. Measure the flux rate with time at various transmembrane pressures. Repeat at several different rates of tangential flow. Typical plots are shown in *Figure 9a*. Determine the optimum transmembrane pressure and tangential flow rate. Predict the expected duration of the final process from the flux rate at the plateau point. Continue the test for this length of time.

5   Continue pumping the solution at the optimum transmembrane pressure and tangential flow rate but do not recycle the permeate. Measure the flux rate. This part of the test will determine the effect of increased concentration on the flux rate (*Figure 9b*).

6   Clean the system by pumping through hot water (e.g. 50 °C; check the manufacturer's recommendations).

7   Measure the flux rate of clean water as in step 1. Ensure that the temperature is identical to that used in step 1. If fouling has occurred the flux rate will not be restored to close to that observed in step 1.

8   Clean the system with one of the solutions suggested by the manufacturer (see Section 3.2.1).

9   Measure the flux rate as in step 1. Unless considerable fouling has occurred this should be ≥ 95% flux rate measured in step 1.

10   Repeat steps 4, 5, 8, and 9 to optimize each parameter (see *Figure 9* for typical results).

11   Finally concentrate several batches under optimum conditions to ensure batch to batch reproducibility. Also measure yields using an assay specific for the protein of interest to determine whether losses are being incurred, e.g. by absorption to the membrane or shear-induced denaturation.

All the plastic membranes are hydrophobic and therefore to prevent irreversible damage they must be stored wet and never allowed to dry out. To prevent microbial contamination the membranes should be stored in the presence of a bacteriostat according to the manufacturer's recommendation (e.g. 0.1% sodium azide or 10% ethanol). This should normally be removed prior to use.

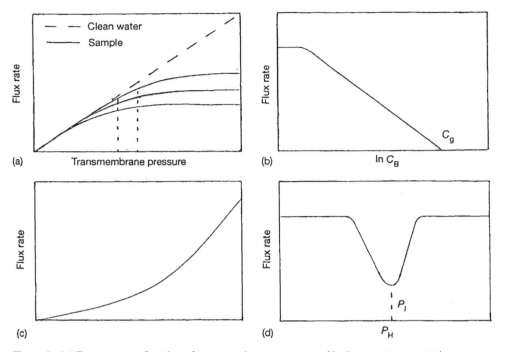

Figure 9; (a) Flux rate as a function of transmembrane pressure with clean water or sample. a, b, and c are obtained at decreasing rates of tangential flow. Above a, the optimum rate of tangential flow, further increases do not increase the flux rate. p, optimum transmembrane pressure. (b) Flux rate as a function of the protein concentration in solution (C_B). Extrapolation to C_g gives the concentration of protein in the gel layer. For diafiltration the optimum protein concentration in solution is $C_g/2.7$. (c) Flux rate as a function of temperature. In practice the optimum temperature is limited to $\leq 40\,°C$ by the stability of the protein. (d) Flux rate as a function of the pH of the solution. At a pH approximately equal to the pI of the protein a minimum flux rate is usually obtained due to precipitation of the protein.

3.2.2 Operating procedures

The reader is referred to the operating instructions supplied by the manufacturer for details of how to set up the system, and what operating pressures to use, etc.

3.3 Other applications of ultrafiltration

3.3.1 Diafiltration

In addition to concentration, ultrafiltration can be used to remove salts from protein solutions (diafiltration). Prior to diafiltration the protein solution is usually concentrated by ultrafiltration. Water or buffer is then added to the retentate and ultrafiltration continued until the filtrate reaches the desired ionic strength and pH. This technique will be described in more detail in a later section in this chapter.

127

3.3.2 Purification

(a) **Separation by size**. Ultrafiltration can to some extent be used to purify proteins on the basis of size (16). Although in principle this appears to be an attractive method of purification, in practice the resolution of the technique is poor due to the following causes.

(i) The distribution of pore sizes is not tight, resulting in some proteins larger than the NMWC passing through the membrane. Since a tenfold change in pore size results in a 100-fold change in NMWC a small change in pore size results in significantly larger proteins passing through.

(ii) Linear molecules pass more readily through the membrane than globular molecules. Thus, linear proteins larger than that predicted by the NMWC can be deformed and squeezed through the pores. In addition pH, ionic strength, and the presence of polyelectrolytes influence the effective size of the protein, since a charged protein has an effectively larger size.

(iii) Concentration polarization effectively lowers the NMWC of the membrane, thus inhibiting the passage of molecules smaller than the NMWC.

(iv) Fouling also causes a lowering of the effective NMWC.

In practice the proteins to be separated should differ in molecular weight by a factor of 100 before purification by ultrafiltration can be used effectively.

(b) **Depyrogenation**. A frequently used application of ultrafiltration is depyrogenation of solutions, such as water for preparation of injectables, antibiotic, and sugar solutions. Most pyrogens are bacterial lipopolysaccharides (LPS), which range in size from subunits of 20 000 molecular weight to aggregates of greater than 0.1 μm in diameter. Ultrafiltration membranes with NMWCs of 10 000 can be used economically and effectively to depyrogenate the filtrate. Thus, using these conditions only molecules with molecular weights less than 10 000 can be depyrogenated. This technique is therefore of limited application for proteins, but could be used for small molecular weight peptides (say < 80 amino acids long). Addition of calcium and/or magnesium ions can induce aggregation of LPS allowing use of membranes with 0.025 μm pores to retain LPS and allow smaller molecules to pass through. For depyrogenating water, reverse osmosis membranes are more commonly used.

(c) **Affinity purification**. Ultrafiltration can be exploited to achieve affinity purification by using a large molecular weight affinity ligand which is retained by the ultrafiltration membrane (17). The protein to be purified will therefore be retained by the membrane, even though it is smaller than the NMWC. The ligand, such as Cibacron blue or p-aminobenzamidine, is covalently coupled to a large molecular weight polymer, such as dextran or starch. The protein mixture to be purified is mixed with the ligand–polymer complex and then ultrafiltered. The protein of interest plus ligand–polymer are retained and can then be separated by dissociation and further ultrafiltration.

A further advance in filtration is the development of membranes for ion exchange or affinity purification (18, 19). In this case the membrane is modified by covalent attachment of an ion exchange group or an affinity ligand. These membranes offer the advantage of higher flow rates over conventional column matrices.

4 Freeze-drying or lyophilization

In contrast to ultrafiltration, lyophilization also results in concentration of any salts present in the initial solution; in addition lyophilization may cause greater losses in enzyme activity. Lyophilization is, however, an invaluable method both for concentrating small molecular weight peptides which are not retained by ultrafiltration membranes, and for obtaining a dry powder of protein. Once obtained, a dry powder of enzyme is more stable than an aqueous preparation of enzyme, since many degradation processes require the presence of water. Hence many commercially available proteins are obtained as freeze-dried powders.

For laboratory scale operations, the solution to be lyophilized is placed in a freeze-drying flask and 'shell-frozen' by slowly rotating the flask in a bath of dry ice and methanol, or liquid nitrogen (NB. These solutions are extremely cold and will cause frost-bite); this results in a film of frozen solution around the outside of the flask. Freeze-drying flasks come in a variety of shapes, pear-shaped flasks are more convenient for small volumes, whilst flat-bottomed flasks are best used for larger volumes. The freeze-drying flask is then rapidly attached to a mechanical vacuum pump (e.g. Edwards), ensuring that all the solution remains frozen prior to applying the vacuum. Any thawed liquid will rapidly degas and 'bump' with possible loss of solution out of the flask. To preserve the life of the pump a cold-trap must be placed between the pump and the frozen solution. This traps the water drawn off by the vacuum and prevents it entering the pump and causing rusting. A simple cold-trap consists of a glass vessel placed in a solution of dry-ice and methanol. Alternatively, purpose built freeze-driers are available from several manufacturers (e.g. Heto); these consist of a cold-trap which is kept at −60 °C by a compressor, and an adaptor for either attaching several freeze-drying flasks, or using several vials. As water is removed from the solution the outside of the flask will become cold, due to evaporation; this effect also ensures that the solution remains frozen. When the outside of the flask warms up to room temperature this indicates that the freeze-drying process is completed; if a vacuum gauge is fitted into the system (e.g. when a purpose built freeze-dryer is used) another indication that the freeze-drying process is completed is given when the gauge reaches a minimum value (approximately equal to the reading obtained from the pump when run in isolation).

For larger scale freeze-drying, purpose designed automated freeze-driers are available. The product is placed on shallow trays, either loose or in vials. Freezing is often done within the freeze-drier prior to application of the vacuum. The shelves are heated during the drying cycle to speed up the process.

If the solution thaws out during the process this may result in greater losses of activity, and will also result in a glassy residue which is difficult to redissolve rather than the light, fluffy powder usually obtained by freeze-drying. Many buffers are suitable for lyophilization. However, phosphate buffers are not ideal since the pH will drop on freezing with subsequent denaturation of the protein. Also buffers with one volatile component should be avoided since again the pH may change dramatically during the lyophilization and on redissolving. Volatile buffers such as ammonium bicarbonate, or water may be preferable to minimize interference with subsequent steps. Buffer concentrations should be minimized to prevent losses in recovery of activity. Additives, such as lactose, trehalose, or mannitol may be added to aid recovery of activity; suitable concentrations are 1–5%. Solutions containing azide should not be lyophilized, particularly using equipment fitted with a condenser, since the equipment becomes potentially explosive.

5 Removal of salts and exchange of buffer

5.1 Dialysis

Frequently it is necessary to remove salts or change the buffer after one step in the purification for the next step to work efficiently (e.g. for ion exchange chromatography, the pH and/or the ionic strength may need to be changed to ensure that the protein will bind to the matrix). This is often achieved by dialysis; the protein solution is placed in a bag of semi-permeable membrane and placed in the required buffer, small molecules can pass freely across the membrane whilst large molecules are retained. The semi-permeable dialysis tubing is usually made of cellulose acetate, with pores of between 1–20 nm in diameter. The size of these pores determine the minimum molecular weight of molecules which will be retained by the membrane (NMWC). Dialysis tubing often requires pre-treatment to ensure a more uniform pore size and removal of heavy metal contaminants; a method for pre-treatment is given in *Protocol 5*. Some dialysis membranes only require wetting in an appropriate buffer and do not require pre-treatment, except for the most sensitive enzymes. Traditional visking tubing has a NMWC of about 15 000–20 000.

Protocol 5

Dialysis

Equipment and reagents
- Dialysis tubing
- Dialysis bag
- Magnetic bar and stirrer motor
- 2% sodium bicarbonate
- 0.05% EDTA
- 20% ethanol
- 0.1% sodium azide

Method

1 Select dialysis tubing of suitable diameter and cut into suitable lengths to contain the volume required.

2 Submerge in a solution of 2% sodium bicarbonate and 0.05% EDTA. Ensure sufficient volume is used to amply cover the dialysis tubing. Boil for 10 min. Ensure the dialysis tubing remains submerged by placing a conical flask partially filled with water on top of the tubing.

3 Discard the solution and boil for 10 min in distilled water. Repeat once more.

4 Cool and place into a suitable solution to prevent microbial growth (e.g. 20% (v/v) ethanol or 0.1% sodium azide). Store at 4°C for up to three months.

Note: Wear gloves for the following steps.

5 Prior to use rinse the dialysis tubing inside and outside with distilled water or buffer.

6 Seal one end of the tubing with a double knot or dialysis clip (e.g. Fisher). Dialysis clips are easy to use, allow easy labelling of each dialysis bag, and float.

7 Pour in the solution to be dialysed (a small funnel may be helpful). It is a good idea to support the bottom end of the bag on the bench. Do not overfill the bag since the volume may increase during dialysis of solutions with higher osmolarity than that used for the dialysis. For solutions of high osmolarity (e.g. 4 M $MgCl_2$ or supernatants from ammonium sulfate precipitations) the volume can increase by twofold or more.

8 Expel the air from the bag and seal the top end with a double knot or dialysis clip.

9 Place the bag in a large volume of buffer and agitate gently with a magnetic bar and stirrer motor. Ensure the bag is not knocked by the magnetic bar, to prevent rupture.

10 Leave to reach equilibrium, usually $\geq$ 3 h, preferably at 4°C.

Small molecules pass freely through the membrane until the osmotic pressure is equalized, thus complete exchange from one buffer requires several changes of the dialysis buffer. For example, if a 20 ml sample containing 1 M NaCl is placed in 1 litre of water the concentration of salt at equilibrium will be:

$$20/1020 \times 1 = 19.6 \text{ M}$$

After changing into a further 1 litre of buffer the concentration would reach 384 nM. Equilibrium is usually reached after approx. 3 h with efficient stirring and 15 000 NMWC membranes; the time taken increases with decreasing NMWC. Dialysis is often carried out overnight, usually at 4°C to minimize losses in activity (*Protocol 5*).

5.2 Diafiltration

A quicker, alternative method for desalting or buffer exchange is diafiltration. This method is also more applicable to larger scale applications (i.e. > 100 ml). Ultrafiltration equipment is used for diafiltration (see Section 3). With some types of equipment, water or buffer is added to the protein solution, which is then concentrated by ultrafiltration; this process is repeated until the ionic strength of the filtrate reaches that of the added buffer or water. With other types or equipment (e.g. Millipore's hollow fibre systems) the plumbing is altered to allow uptake of water or buffer at a rate equal to the flux through the membrane; this method has the advantage that it can be left unattended whilst equilibrium is achieved. The time taken to achieve equilibrium depends on the volume of sample and the flux rate (as an indication 1 litre can be diafiltered in 2–4 h). The optimum concentration for diafiltration can be determined as described in *Protocol 4* (*Figure 9b*). Higher concentrations will decrease the flux rate and hence increase the processing time, whilst lower concentrations and hence larger volumes will require larger volumes of buffer. The reader is referred to the instructions supplied by the system manufacturer for details on how to carry out diafiltration.

5.3 Gel filtration

Another quicker alternative to dialysis is gel filtration. This method is only applicable to small volumes. The maximum sample volume should not exceed 25–30% of the volume of the column to ensure adequate resolution between the protein and salt. A gel filtration matrix with a small pore size (e.g. Sephadex G-25 Amersham Pharmacia) is poured into a column to give a bed volume of approximately five times the volume of sample to be desalted. A syringe plugged with glass wool or a glass fibre disc can be used, or small disposable columns (e.g. Bio-Rad's Econo-Pac). Pre-packed columns for desalting and buffer exchange are available from Amersham Pharmacia (PD-10 columns). Methods for using these columns are given in *Protocol 6*. Unfortunately, this method of desalting and buffer exchange results in dilution of the sample.

6 Purification and concentration by precipitation

Many of the early protein purification procedures used only precipitation methods as a means of separating one protein (or class of proteins) from another. For example, the core histones (H2a, H2b, H3, and H4) were purified by ethanol and/or acetone precipitation (20). Differences in solubility have also been used to purify albumins and globulins from serum (21) (the globulins are precipitated by diluting serum with water, whilst the albumins remain soluble). An example of purification by precipitation is given in *Protein purification applications: a practical approach* (22), for the purification of mammalian cytochrome oxidase. Nowadays precipitation is usually only used as a fairly crude separation step often during the early stages of a purification procedure, and this is then followed by chromatographic separations. Precipitation can also be used as a method of concentrating proteins prior to analysis or a subsequent purification step.

The solubility of a protein molecule in an aqueous solvent is determined by the distribution of charged hydrophilic and hydrophobic groups on its surface. The charged groups on the surface will interact with ionic groups in the solution (*Figure 10*). Protein precipitates are formed by aggregation of the protein molecules, induced by changing pH or ionic strength, or by addition of organic miscible solvents or other inert solutes or polymers. Temperature will also affect the degree of aggregation achieved. Precipitates can be recovered by filtration or centrifugation, washed, and redissolved in an appropriate buffer, if required. To remove final traces of the precipitating agent, which might interfere with subsequent purification steps, the redissolved precipitate should be dialysed, diafiltered, or desalted on a gel filtration column (see Section 5).

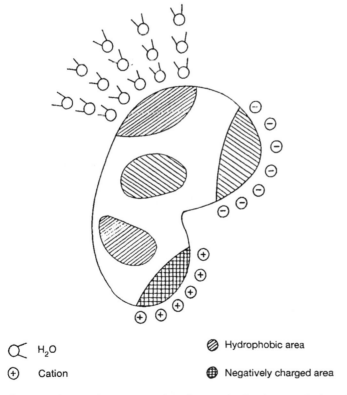

⋉	H₂O	⬀ Hydrophobic area	
⊕	Cation	⬢ Negatively charged area	

Figure 10 Schematic representation of a protein showing negatively and positively charged areas on the protein interacting with ions in the solution. The hydrophobic areas on the protein interact with water molecules causing an ordered matrix of water molecules to form over these areas.

Protocol 6

Desalting columns

Equipment and reagents

- Centrifuge
- Chromatography column or plastic syringe
- Glass wool
- Sephadex G-25
- Buffer

Method

1 Select a suitable size column. This may be a chromatography column or a plastic syringe plugged with a small amount of glass wool.

2 Pack the pre-swollen matrix (e.g. Sephadex G-25) into the column (the length of the column should be > twice its diameter to ensure adequate resolution). To do this make a slurry, pour into the column, and allow to settle. If required top up the volume with additional slurry. (A perfectly packed column is not necessary since high resolution is not required.)

3 To equilibrate the column in the required buffer pass more than a column volume through by gravity or using a pump (if using a pump ensure the maximum recommended flow rate is not exceeded). Alternatively, matrices packed in syringes can be centrifuged as follows. Place the syringe in a tube, supported off the bottom of the tube by the flanges on the syringe. Place approximately a column volume of buffer on top of the matrix and centrifuge at 1600 g for 4 min.

4 Place the sample to be desalted on top of the matrix and allow it to pass into the matrix by the method used in step 3. The volume of sample must be ≤ 30% of the volume of matrix. Discard the flow-through.

5 Continue passing through buffer until one column volume has been used and collect suitable size fractions (e.g. one-tenth of the total column volume). Larger fractions will cause higher dilution of the sample, and will decrease the resolution.

6 Determine which fractions contain the protein by measuring absorbance at 280 nm or by using another protein assay. The protein will usually elute between 0.2–0.7 times the column volume.

6.1 Precipitation by alteration of the pH

One of the easiest methods of precipitating a protein and achieving a degree of purification is by adjusting the pH of the solution to close or equal to the pI of the protein (termed isoelectric precipitation). The surface of protein molecules is covered by both negatively and positively charged groups. Above the pI the surface is predominantly negatively charged, and therefore like-charged molecules will be repelled from one another; conversely below the pI the overall charge will be positive and again like-charged molecules will repel one another. However, at the pI of the protein the negative and positive charges on the surface of a molecule cancel one another out, electrostatic repulsion between individual molecules no longer occurs, and electrostatic attraction between molecules may occur, resulting in formation of a precipitate.

Isoelectric precipitation is often used to precipitate unwanted proteins, rather than to precipitate the protein of interest, since denaturation and inactivation can occur on precipitation (see Section 6.6).

6.2 Precipitation by decreasing the ionic strength

Some proteins can be precipitated by lowering the ionic strength. This can rarely be achieved with crude extracts, since the ionic strength can only be lowered by addition of water, which will also lead to a decrease in the concentration, and hence an increased solubility (a notable exception is the serum globulins). However, this form of precipitation can often occur at later stages of a purification, for example, when removing salts by diafiltration, dialysis, or gel filtration. This may not always be a welcome occurrence, for example, when using gel filtration a precipitated protein will be trapped and may block the matrix. Precipitation at low ionic strength is more likely to occur at or close to the pI of the protein, since the causes of precipitation are similar and therefore additive.

6.3 Precipitation by increasing the ionic strength (salting-out)

Precipitation by addition of neutral salts is probably the most commonly used method for fractionating proteins by precipitation. The precipitated protein is usually not denatured and activity is recovered upon redissolving the pellet. In addition these salts can stabilize proteins against denaturation, proteolysis, or bacterial contamination. Thus, a salting-out step is an ideal step at which to store an extract overnight, either before or after centrifugation. The cause of precipitation is different from that for isoelectric precipitation, and therefore the two are often used sequentially to obtain differential purification. Salting-out is dependent on the hydrophobic nature of the surface of the protein. Hydrophobic groups predominate in the interior of the protein, but some are located at the surface, often in patches. Water is forced into contact with these groups, and in so doing becomes ordered (*Figure 10*). When salts are added to the system, water solvates the salt ions and as the salt concentration increases water is removed from around the protein, eventually exposing the hydrophobic patches. Hydrophobic patches on one protein molecule can interact with those on another, resulting in aggregation. Thus, proteins with larger or more hydrophobic patches will aggregate and precipitate before those with smaller or fewer patches, resulting in fractionation. The aggregates formed are a mixture of several proteins, and like isoelectric precipitation the nature of the extract will affect the concentration of salt required to precipitate the protein of interest. In contrast to isoelectric precipitation, increasing the temperature increases the amount of precipitation; however, salting-out is usually performed at 4°C to decrease the risk of inactivation (by, e.g. proteases).

Several aspects of the salt used should be considered. The effectiveness of the salt is mainly determined by the nature of the anion, multi-charged anions being the most effective; the order of effectiveness is phosphate > sulfate > acetate > chloride > (and follows the Hofmeister series). Although phosphate is more effective than sulfate, in practice phosphate consists of mainly HPO_4^{2-} and $H_2PO_4^-$ ions at neutral pH, rather than the more effective PO_4^{3-}. Monovalent cations are most effective, with $NH^{4+} > K^+ > Na^+$. The salt must be relatively cheap with few impurities present. The solubility is also an important consideration, since concentrations of several molar are required; thus, many potassium salts are not suitable. Because of the risk of possible denaturation, or changes in solubility, there should be little increase in heat caused by the salt dissolving. The final consideration is the density of the resultant solution, since the difference between the densities of the aggregate and the solution determines the ease of separation by centrifugation.

In practice ammonium sulfate is the most commonly used salt (other salts which have been used in particular applications are ammonium acetate, sodium sulfate, and sodium citrate). Ammonium sulfate is cheap, and sufficiently soluble; a saturated ammonium sulfate solution in pure water is approximately 4 M. The density of a saturated solution is 1.235 g/ml, compared to 1.29 g/ml for a protein aggregate in this solution. In practice the density of a 75–100% saturated solution may be higher than 1.235 g/ml, due to the presence of other salts and compounds in the extract, therefore making recovery of a protein aggregate by centrifugation difficult.

The ammonium sulfate concentration is usually quoted as per cent saturation, assuming that the extract will dissolve the same amount of ammonium sulfate as pure water. To calculate the number of grams of ammonium sulfate (g) to add to 1 litre at 20°C to give a desired concentration use the following equation:

$$g = \frac{533 \, (S_2 - S_1)}{100 - 0.3 \, S_2}$$

where S_1 is the starting concentration, and S_2 is the final concentration. This equation allows for the increase in volume that occurs on addition of the salt. Alternatively, the amount of ammonium sulfate to add can be read from *Table 2*.

Ammonium sulfate will slightly acidify the extract, therefore a buffer of about 50 mM should be used to maintain a pH between 6.0–7.5. If a higher pH is preferred then sodium citrate should be used instead of ammonium sulfate. Although ammonium sulfate is sufficiently pure for most applications, if the enzyme of interest is sensitive to heavy metals, EDTA should be included in the buffer (even if the level of contamination is only 1 part per million, with an ammonium sulfate concentration of 75% saturation, the concentration of contaminant will be 3 μM). Usually an ammonium sulfate cut is taken in order to obtain a higher degree of purification. Thus, the extract is taken to a per cent saturation where the protein of interest does not precipitate and the precipitate is removed by centrifugation. More ammonium sulfate is added to the supernatant

Table 2 The amount of solid ammonium sulfate to be added to a solution to give the desired final saturation at 0 °C

	Final concentration of ammonium sulfate, % saturation at 0 °C																
	20	25	30	35	40	45	50	55	60	65	70	75	80	85	90	95	100
Initial concentration of ammonium sulfate																	
	g solid ammonium sulfate to add to 100 ml of solution																
0	10.7	13.6	16.6	19.7	22.9	26.2	29.5	33.1	36.6	40.4	44.2	48.3	52.3	56.7	61.1	65.9	70.7
5	8.0	10.9	13.9	16.8	20.0	23.2	26.6	30.0	33.6	37.3	41.1	45.0	49.1	53.3	57.8	62.4	67.1
10	5.4	8.2	11.1	14.1	17.1	20.3	23.6	27.0	30.5	34.2	37.9	41.8	45.8	50.0	54.5	58.9	63.6
15	2.6	5.5	8.3	11.3	14.3	17.4	20.7	24.0	27.5	31.0	34.8	38.6	42.6	46.6	51.0	55.5	60.0
20	0	2.7	5.6	8.4	11.5	14.5	17.7	21.0	24.4	28.0	31.6	35.4	39.2	43.3	47.6	51.9	56.5
25		0	2.7	5.7	8.5	11.7	14.8	18.2	21.4	24.8	28.4	32.1	36.0	40.1	44.2	48.5	52.9
30			0	2.8	5.7	8.7	11.9	15.0	18.4	21.7	25.3	28.9	32.8	36.7	40.8	45.1	49.5
35				0	2.8	5.8	8.8	12.0	15.3	18.7	22.1	25.8	29.5	33.4	37.4	41.6	45.9
40					0	2.9	5.9	9.0	12.2	15.5	19.0	22.5	26.2	30.0	34.0	38.1	42.4
45						0	2.9	6.0	9.1	12.5	15.8	19.3	22.9	26.7	30.6	34.7	38.8
50							0	3.0	6.1	9.3	12.7	16.1	19.7	23.3	27.2	31.2	35.3
55								0	3.0	6.2	9.4	12.9	16.3	20.0	23.8	27.7	31.7
60									0	3.1	6.3	9.6	13.1	16.6	20.4	24.2	28.3
65										0	3.1	6.4	9.8	13.4	17.0	20.8	24.7
70											0	3.2	6.6	10.0	13.6	17.3	21.2
75												0	3.2	6.7	10.2	13.9	17.6
80													0	3.3	6.8	10.4	14.1
85														0	3.4	6.9	10.6
90															0	3.4	7.1
95																0	3.5
100																	0

to give a per cent saturation where the protein of interest does precipitate, and can therefore be obtained by centrifugation. In order to determine the appropriate per cent saturations follow the method given in *Protocol 7*. As mentioned before the composition of the extract will influence the precipitation of the protein, as will its concentration. These initial trial studies must therefore be carried out on an extract obtained by the same conditions that will be used in the purification procedure. The temperature at which the precipitation is carried out is also important; the higher the temperature the lower the solubility of the protein. Slow addition of the ammonium sulfate and efficient stirring are important, particularly as the desired saturation is approached. Dissolved air may come out of solution and cause frothing; this is not deleterious, but frothing caused by over-vigorous stirring may cause denaturation of the protein. The precipitate is usually removed by centrifugation, though filtration can be used, particularly if the density of the protein aggregate is similar to or lower than that of the solution. For centrifugation 100 000 g/min is normally sufficient, although the higher salt concentrations may require more; thus, 10 000 g for 10 min or 3000 g for 35 min may be used. After centrifugation the pellet should dissolve in a volume of buffer equal to one or two times the volume of the pellet; any material which does not dissolve is particulate or denatured protein, and should be removed by centrifugation. The redissolved pellets will contain ammonium sulfate, therefore any assays used should not be sensitive to ammonium sulfate; the Bradford dye-binding assay is a suitable assay for total protein. Desalting (Section 5) prior to assaying may be necessary. The choice of appropriate ammonium sulfate concentrations for the cut will be a compromise between the yield and the degree of purification; thus, a narrow cut will result in a lower yield with a higher degree of purification.

Protocol 7

Optimization of ammonium sulfate precipitation

1 Take aliquots of the extract, ideally 20–50 ml, place in beakers, and pre-chill to 4 °C.

2 Calculate the amount of ammonium sulfate required to give 20, 30, 40, 50, 60, 70, and 80% saturation from the equation given in the text or *Table 2*. Weigh out the required amounts of ammonium sulfate and ensure all lumps are removed (use a pestle and mortar).

3 Slowly add the ammonium sulfate to each aliquot whilst stirring (use a magnetic follower and stirrer). Leave each aliquot stirring for 1 h at 4 °C.

4 Centrifuge each aliquot at 3000 g for 40 min.

5 Remove the supernatants and drain the pellets. Dissolve the pellets in buffer (e.g. PBS or 50 mM Tris–HCl pH 8). Use the same volume for each pellet, approximately twice the volume of the largest pellet.

6 If undissolved material remains, centrifuge at 3000 g for 15 min.

7 Assay the supernatants for total protein using the Bradford assay and by an assay specific for the protein of interest. (Dialyse the supernatants if ammonium sulphate interferes with the assay.)

8 Plot the concentration of total protein and the protein of interest against % saturation of ammonium sulfate. For an ammonium sulfate cut choose the maximum % saturation which does not precipitate the protein of interest, and the minimum % saturation that precipitates all the protein of interest (see *Figure 11*).

9 Repeat the procedure with one aliquot. First add ammonium sulfate to give the lowest % saturation chosen. After centrifuging, add further ammonium sulfate to give the highest % saturation chosen. Check that none of the protein of interest is in the first pellet or in the supernatant from the higher % saturation. Since the extent of the precipitation depends on the nature of the extract, the higher % saturation may require further optimization to achieve maximum purification.

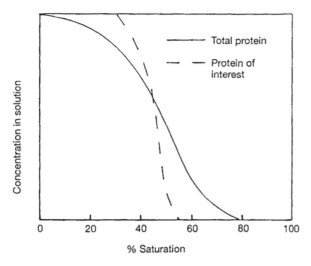

Figure 11 Typical profile for ammonium sulfate precipitation. The optimum % saturations for an ammonium sulfate cut would be 30% and 55% for purification of this protein.

6.4 Precipitation by organic solvents

Many proteins can be precipitated by addition of water-miscible organic solvents, such as acetone and ethanol. The factors which influence the precipitation behaviour of a protein are similar to those involved in isoelectric precipitation and different from those involved in salting-out; thus, this method can be used as an alternative to isoelectric precipitation in a purification sequence, perhaps in conjunction with salting-out. Addition of the organic solvent lowers the dielectric constant of the solution, and hence its solvating power. Thus, the solubility of a protein is decreased and aggregation through electrostatic attraction can occur. Precipitation occurs more readily when the pH is close to the pI of the protein. The size of the protein also influences its precipitation behaviour; thus, a larger protein will precipitate in lower concentrations of organic solvent than a smaller protein with otherwise similar properties. However, some hydrophobic proteins, particularly those which are located in the cellular membranes are not precipitated by organic solvents, and in fact can be solubilized from the membranes by addition of organic solvents (see ref. 23). With these proteins the organic solvent will displace the water molecules from around the hydrophobic patches of the protein, resulting in an increased solubility.

To minimize denaturation, precipitation with organic solvents should be carried out at or below 0 °C. At higher temperatures the protein conformation will be rapidly changing, thus enabling molecules of the organic solvent to gain access to the interior of the protein, where they can disrupt the hydrophobic interactions and cause denaturation. Fortunately mixtures of aqueous solutions and organic solvents freeze at well below 0 °C. However, as the organic solvent is added the temperature of the solution will rise due to the negative $\Delta H°$ of hydration of the solvent molecules. Thus, care must be taken to prevent the temperature rising above 10 °C, by using pre-cooled organic solvent, and adding it slowly, with mixing, to the solution contained in a vessel in an ice-salt bath (or other low temperature bath). Lower temperatures will also promote the degree of precipitation. The ionic strength of the solution should be between 0.05–0.2; higher ionic strengths require higher amounts of organic solvent with an increased risk of denaturation, whilst lower ionic strengths can result in the precipitate being very fine and difficult to sediment. Prior to adjustment of the pH close to the pI of the protein will result in precipitation at a lower concentration of organic solvent.

Acetone and ethanol are the most commonly used solvents; others which have been used are methanol, propan-1-ol, and propan-2-ol. Safety aspects should be considered, particularly when working on a large scale; thus, the solvent should be relatively non-toxic and have a relatively high flashpoint, above 20 °C (thus dioxans and ethers should be avoided). The longer chain alcohols, such as butanol, cause a higher degree of denaturation than ethanol. In many cases acetone is preferable, since lower concentrations are required, and therefore less denaturation occurs. Most proteins will be precipitated by addition of an equal volume of acetone, or four volumes of ethanol. Clearly addition of such large volumes will result in dilution of the protein, thus the starting concentration of protein should be more than 1 mg/ml. The concentration of organic solvent is usually expressed as a percentage, assuming that the volumes are additive. Thus, if an equal volume of acetone is added this corresponds to 50% (v/v); however, there will be a small decrease in volume, approximately 5% with acetone, which is due to the formation of hydrated solvent molecules that occupy a smaller volume than their constituent components.

As with salting-out, an organic solvent cut is often taken to increase the degree of purification; the optimum amount of organic solvent to add can be determined using a similar method to that described in *Protocol 7*. The following equation is used to calculate the volume, in ml, of organic solvent to add to one litre:

$$v = \frac{1000 \, (P_2 - P_1)}{100 - 0.3 \, P_2}$$

where P_1 is the starting percentage of organic solvent, and P_2 is the desired percentage of organic solvent. Since the density of the organic solvent mix is often less than that of water, the precipitates will sediment more readily than, for example, those formed with ammonium sulfate. Thus, the precipitate may be left to settle by gravity and the supernatant decanted, or centrifugation may be used (e.g. 10 000 g for 5 min). If centrifugation is used, the rotor should be pre-chilled and the centrifuge should be maintained at 0 °C, otherwise if the solution is allowed to warm up denaturation may occur, and/or some of the precipitate may redissolve. The resultant precipitate will normally redissolve in one or two times its own volume; any undissolved material will probably be denatured protein and can be removed by centrifugation. Organic solvent will be present in the redissolved pellet, therefore it is important to ensure that this will not interfere with the assays, or any subsequent purification steps. The solvent can be removed by evaporation at reduced pressure, or by dialysis or gel filtration, preferably at 4°C. In the case of subsequent purification steps only those involving hydrophobic interactions will be affected (precipitation with ammonium sulfate is usually possible in the presence of low concentrations of organic solvents, but a higher concentration of salt will be required).

6.5 Precipitation by organic polymers

PEG is the most commonly used organic polymer (24). The mechanism of precipitation is similar to that of precipitation by organic solvents, however, lower concentrations are required, usually below 20%. Higher concentrations result in viscous solutions, making recovery of the precipitate difficult. The molecular weight of the polymer should be greater than 4000; the most commonly used molecular weights are 6000 and 20000. PEG can be removed by ultrafiltration, provided its molecular weight differs significantly from that of the protein of interest. However, PEG does not interfere with many of the possible subsequent purification steps (e.g. ion exchange or affinity chromatography).

6.6 Precipitation by denaturation

Precipitation by denaturation can be used as a purification step if the protein of interest is not denatured by the treatment, whilst many of the contaminant proteins are. This method can also be used to concentrate the proteins in a solution prior to analysis (e.g. for gel electrophoresis or amino acid analysis). Denaturation can be caused by changes in temperature, pH, or addition of organic solvents. The tertiary structure of the proteins is disrupted during denaturation, resulting in the formation of random coil structures. In solution these random coils become entangled with one another, thus forming aggregates. Aggregate formation is influenced by pH and ionic strength, occurring more readily close to the pI of the protein, and at lower ionic strength.

6.6.1 Denaturation by high temperatures

High temperatures induce denaturation by breaking many of the bonds holding the protein in its native conformation (e.g. van der Waals forces, ionic interactions, and even the peptide bond itself at the higher extremes), and also causing the release of associated solvent molecules. Different proteins are denatured at different temperatures, thus achieving purification; for example, human tumour necrosis factor has been partially purified in this way. Small scale trials using 1 ml of extract can be carried out to determine a temperature (usually between 45 and 65°C) and incubation time which will cause maximum precipitation of contaminating proteins, with minimum loss of activity of the protein of interest. For these small scale trials the extract used should be obtained in exactly the same way as it will in the purification itself, since pH, ionic strength, and composition all affect the precipitation obtained. Although on a small scale short times (e.g. 1 min) may be practical, due to the increased time a larger volume will take to reach the temperature, and cool down, a time of about 30 min or longer is usually more practical. Unfortunately many proteases are relatively stable to high temperatures, and can in fact be more active under these conditions, thus protease inhibitors should be added prior to this step.

6.6.2 Denaturation by extremes of pH

Extremes of pH cause internal electrostatic repulsion, or loss of internal electrostatic attraction by changing the charges on the side chains of the amino acids. Thus, the protein then opens up and bound solvent is lost, resulting in denaturation. This method of purification is often extremely useful as a preliminary purification step of recombinant proteins expressed in prokaryotes, since many bacterial proteins have pIs in the region of 5.0 and can therefore often be removed by adjusting the pH of the extract to around 5.0. In general, many proteins can be precipitated by adjusting the pH to 5 or below, fewer proteins precipitate at neutral and alkaline pHs. Before using this treatment as a purification step, the stability of the protein of interest under these conditions should be established. Use of strong acids and bases should be avoided; thus use acetic acid for a pH of 4 or above, or citric acid for a pH of 3 or more and diethanolamine or sodium carbonate for a pH of 8 or more. Small scale trials should be carried out to determine the optimum pH. Remember when determining the optimum pH that the aggregate formed is made up of many proteins, as well as particulate material such as ribosomes and membrane fragments; thus the composition of the aggregate will depend on the composition of the extract. The extract used to determine the optimum pH must therefore be the same as that to be used in the purification itself. Also, if the earlier purification procedures of an established protocol are changed (e.g. the method of extraction) the optimum pH should be redetermined; or the protein of interest might end up in the wrong fraction and be inadvertently discarded! The trials should also be carried out at the same temperature as will be used in the purification; increased temperature will cause increased precipitation.

Strong acids, such as perchloric acid or trichloroacetic acid (TCA), can be used to concentrate proteins for analysis when activity is not important, or occasionally for purification (e.g. perchloric acid is used to extract histones and HMG proteins as a first step in their purification). Care should be taken with these acids, since they are highly corrosive. In addition perchloric acid can form explosive compounds, particularly when in contact with wood. Most proteins will be precipitated by addition of TCA to 10% (v/v); smaller proteins, of molecular weight less than 20 000, may require up to 20%. Excess TCA can be removed from the protein pellets by washing with buffer.

6.6.3 Denaturation by organic solvents

Selective denaturation by organic solvents is often carried out at 20–30 °C to promote denaturation. Small scale trials should be used to establish optimum conditions, remembering that pH, temperature, ionic strength, and extract composition will affect the precipitation behaviour. For analysis when activity is not required, to ensure maximum precipitation, the organic solvent should be added at room temperature and then the mixture cooled to −20°C for at least an hour prior to centrifugation. Four volumes of acetone or nine volumes of ethanol may be required to achieve maximum precipitation.

7 Aqueous two-phase partitioning

B. A. Andrews and J. A. Asenjo

Chemical Engineering Department, University of Chile, Santiago, Chile.

Aqueous two-phase partitioning exploits the incompatibility between aqueous solutions of two polymers, or a polymer and a salt at high ionic strength. This incompatibility arises from the inability of the polymer coils to penetrate into each other. Hence, as the polymers are mixed large aggregates form and the two polymers will tend to separate due to steric exclusion. The most commonly used polymers are PEG and dextran (25, 26). A similar exclusion phenomenon can be observed between a polymer and a high concentration of salt (e.g. PEG and phosphate) (27).

Two-phase partitioning can be used to separate proteins from cell debris, to purify proteins from other proteins, or to concentrate proteins. Most soluble and particulate matter will partition to the lower, more polar phase, whilst proteins will partition to the top, less polar phase (usually PEG). Separation of proteins from one another is achieved by manipulating the partition coefficient by altering the average molecular weight of the polymers, the type of ions in the system, the ionic strength, or the presence of hydrophobic groups (28, 29). To achieve a higher degree of purification several sequential partitioning steps can be carried out, or alternatively polymers can be mixed to yield more than two phases. Affinity partitioning can also be used to increase the degree of purification; in this case affinity ligands, such as triazine dyes (e.g. Cibacron blue) are covalently attached to one or both polymers. Alternatively, one polymer may be modified with hydrophobic groups to alter the partitioning of the protein (26). Two-phase aqueous partitioning is a very mild method of protein purification, and denaturation or loss of biological activity are not usually seen. This is probably due to the high water content and low interfacial tension of the systems which will therefore protect the proteins. The polymers themselves may also have a stabilizing effect.

Phase diagrams can be constructed for each polymer system and are used to predict the volume and compositions of the top phases. The curve T–C–B in *Figure 12* is called the binodial curve. It separates the area to the left of TCB where there is only one phase present, with both PEG and dextran in solution, and the area to the right of TCB where two phases form. The overall com-

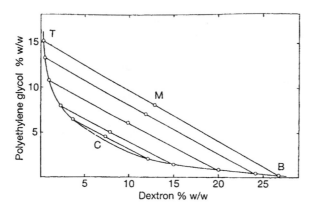

Figure 12 Phase diagram and phase compositions of the dextran/PEG system D24–PEG

position of any particular system is given by the ordinate (PEG) and the abcissa (dextran) values. For the region to the left of TCB (one phase) this is straightforward. For the region to the right of TCB (two phases) a mixture of composition given by point M will separate into two phases of composition T (top) and B (bottom), where T and B are on the equilibrium tie line through point M and on the binodial curve. Other tie lines are also shown. Tie lines are defined as the straight lines that join the composition values of two phases in equilibrium. They will be on different sides of the binodial curve. Tie lines will never cross each other. All points on one tie line will have identical compositions in the top (T) and bottom (B) phases, but the relative phase volumes and hence overall composition will be different. The ratio of the volumes of the upper to lower phases is given by the ratio of the length of the line sections MB/MT. Thus, to concentrate a protein the phase into which the protein partitions should have a small relative volume (i.e. M will be nearer the bottom of the tie line). At the critical point (C) the two phases theoretically have identical compositions, are of equal volumes and have a partition coefficient of 1.

The partition coefficient (K) is defined as:

$$K = C_T/C_B$$

where C_T and C_B represent the equilibrium concentrations of the partitioned protein in the top and bottom phases, respectively. Systems represented by points on the same tie line have identical partition coefficients.

The mechanism governing partition is largely unknown; qualitatively it can be described as follows (26). A protein interacts with the surrounding molecules within a phase via various bonds, such as hydrogen, ionic, and hydrophobic interactions, together with other weak forces. The net effect of these interactions is likely to be different in the two phases and therefore the protein will partition into one phase where the energy is more favourable. If the energy needed to move a protein from one phase to the other is ΔE, at equilibrium the relationship between the partition coefficient (K) and ΔE can be expressed as:

$$K = \exp(\lambda E/kT)$$

where k is the Boltzman constant and T the absolute temperature. E will depend on the size of the partitioned molecule, since the larger it is, the greater the number of atoms exposed which can interact with the surrounding phase. Thus, the Bronsted theory gives an exponential relationship between the partition coefficient and molecular weight (M):

$$K = \exp(\lambda M/kT)$$

where λ is a factor which depends on properties other than molecular weight, such as charge and density. The amount of protein extracted into the top phase (Y_t) can be determined using the following equation:

$$Y_t\,(\%) = \frac{100}{1 + V_b/V_t.K}$$

where V_t and V_b are the volumes of the top and bottom phases, respectively.

The following properties of partitioning can be exploited individually or in conjunction to achieve an effective separation of a particular protein (26):

(a) Size-dependent partition where molecular size of the proteins or surface area of the particles is the dominating factor.

(b) Electrochemical, where the electrical potential between the phases is used to separate molecules or particles according to their charge.

(c) Hydrophobic affinity, where the hydrophobic properties of a phase system are used for separation according to the hydrophobicity of proteins.

(d) Biospecific affinity, where the affinity between sites on the proteins and ligands attached to one of the phase polymers is exploited for separation.

(e) Conformation-dependent, where the conformation of the proteins is the determining factor.

Thus, the overall partition coefficient can be expressed in terms of all these individual factors:

$$\ln K = \ln K_0 + \ln K_{el} + \ln K_{hfob} + \ln K_{biosp} + \ln K_{size} + \ln K_{conf}$$

where el, hfob, biosp, size, and conf stand for electrochemical, hydrophobic, biospecific, size, and conformational contributions to the partition coefficient, and $\ln K_0$ includes other factors.

The factors which influence partitioning of a protein in aqueous two-phase systems as follows:

(a) The polymer used.

(b) Molecular weights and size of polymers.

(c) Concentration of polymer.

(d) Ionic strength.

(e) pH.

(f) Purity of protein solution.

Generally, the higher the molecular weight of the polymers the lower the concentration needed for the formation of two phases, and the larger the difference in molecular weights between the polymers the more asymmetrical is the curve of the phase diagram. Also, the larger the molecular weight of the PEG, the lower the value of K, whereas the molecular weight of dextran does not have such a strong effect on K. A high pH can also result in a several-fold increase in the value of K (27), as can a high phosphate concentration (e.g. 0.3–0.4 M) (25).

7.1 Equipment and materials

Polymers used for two-phase partitioning are listed in *Table 3*, together with their properties. For large scale use the highly purified dextrans are prohibitively expensive; thus, a high molecular weight (5 000 000) crude dextran and a crude dextran after limited acid hydrolysis have been used (30). Starch polymers have also been used as substitutes for dextran (31).

At a laboratory scale, equipment will mainly consist of plastic centrifuge tubes provided with lids, pipettes, a centrifuge, and a small ultrafiltration unit with a low molecular weight cut-off membrane (e.g. 2000). For a large scale operation usually a liquid–liquid separator (e.g. nozzle centrifuge), or a mixer settler will be required. This will mainly depend on the density difference between the two phases and their viscosity. Generally, in a PEG/phosphate system more rapid separation is obtained than in a PEG/dextran (or a PEG/starch polymer) system (10–60 min as opposed to 30–180 min using gravitational settling).

7.2 Optimization

To develop a suitable two-phase system for the concentration of a specific protein all the factors which can influence the system must be taken into account. These parameters will also influence each other (27).

From the phase diagram a system can be constructed to allow concentration of the protein into the top phase. For example, using *Figure 13*, a system with an overall composition at point M would have a large potassium phosphate bottom phase and a small PEG-rich top phase into which the protein will partition.

When developing methods at the laboratory scale 1–10 ml systems are used (32). The method given below is suitable for this scale.

(a) Prepare stock solutions on a weight to weight basis in water (e.g. 20–30% (w/w) for dextran or Ficoll, 30–40% (w/w) for PEG, and for salts at least four

Table 3 Polymers used in two-phase systems

Polymer	Composition	Molecular weight
Polyethylene glycol (PEG)	Polymer of ethylene glycol	600–20 000; usually 4000–8000
Dextran, technical grade	(1–6) linked glucose polymer	500 000
Ficoll	Co-polymer of sucrose and epichlorohydrin	400 000
Reppal PES	Hydroxypropyl derivative of starch	100 000 or 200 000
Aquaphase PPT	Hydroxypropyl derivative of starch	

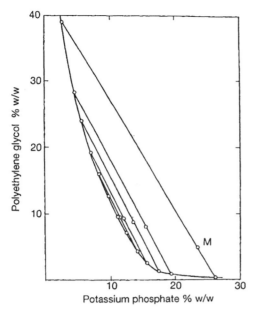

Figure 13 Phase diagram and phase compositions of the potassium phosphate–PEG 4000 system at 20 °C (26).

times the concentration required in the phase system). Add sodium azide to 0.05% if the solutions are to be stored for more than one week. Store at 4 °C.

(b) Weigh the required amounts of polymer stock solutions in graduated glass tubes. Ensure the temperature of the stock solutions has been equilibrated to that at which the partitioning will be carried out (usually ~ 20 °C).

(c) Add the sample and water to give the appropriate concentration of polymers (e.g. a system of 15% PEG/15% potassium sulfate contains the given % (w/w) in the final overall composition of the system).

(d) Mix for ~ 30 sec on a vortex mixer.

(e) Separate the two phases by either:

   (i) Centrifuging at 100–2000 g for 5 min using a swing-out rotor.

   (ii) Gravitational settling, typically for 10 min to 2 h. The settling time varies with the system type, the volumes used, and the ratio of the volumes. Also sedimentation time decreases with increasing length of the tie line due to higher interfacial tension, and larger density differences (thus, systems close to the critical point require longer settling times). With increasing polymer concentration this effect may be counteracted by the increase in viscosity (26).

(f) Measure the volumes of the two phases and remove samples from both with a Pasteur pipette.

(g) Measure the total protein concentration and/or enzyme activity in each phase. Ensure that the assay is not affected by the polymers (e.g. PEG strongly affects the Folin protein assay).

147

At the laboratory scale the following parameters should be determined to assist optimization (32):

- concentration of protein of interest in both phases
- volume of each phase
- material balance of protein of interest
- concentration of contaminants in each phase (e.g. total protein, nucleic acids)
- material balance of contaminants
- pH of each phase

These parameters will be influenced by a variety of factors (33–36), such as polymer molecular weight and concentration, pH, ionic strength, and salt type. Thus, for each purification, these factors should be optimized.

The following general rules can be applied to optimize partition. To increase the partition coefficient of a protein in the PEG phase of PEG/dextran systems:

(a) Lower the molecular weight of PEG.

(b) Increase the molecular weight of dextran.

(c) Increase the pH in the presence of phosphate.

(d) Increase the phosphate concentration.

If the pI of the protein is known then the influence of salt type (used to change the electrical potential difference between both phases) on the partition coefficient can be more readily predicted.

For separating insoluble material, such as cell debris, from proteins by partition, conditions must be established so that solids are collected in the bottom phase and the protein in the top phase (with an enhanced enrichment). A screening for suitable phase systems may start with the following system compositions: 9% PEG 4000; 2% dextran 500 (or 1% crude dextran); 0.2 M potassium phosphate pH 7–8; biomass content approximately 20% (wet weight/v). PEG and dextran concentrations may then be varied in both directions and the partitioning of the protein should be optimized as described above.

7.3 Using phase partitioning in a purification procedure

Partitioning using aqueous systems will be predominantly used for cell debris removal and the primary purification steps. As a method of cell debris removal, aqueous phase systems offer a number of important advantages (36). In addition to cell debris removal, the protein of interest is enriched and other contaminants, such as proteins, nucleic acids, polysaccharides, and coloured by-products from fermentation can also be removed. The polymers present in the aqueous phases also enhance stability of the protein and little or no protein denaturation is observed. The process can also be carried out at room temperature without cooling.

For economic reasons the amount of cell mass processed within the phase system should be as high as possible. However, with increased cell mass the

Table 4 Extraction of enzymes from cell homogenates (30)

Organism	Enzyme	Constituent of the phase system	K_{enzyme}	Yield (%)
Candida boidinii	Catalase	PEG 4000/crude dextran	2.95	81
	Formaldehyde dehydrogenase	PEG 4000/crude dextran	10.8	94
	Formate dehydrogenase	PEG 4000/crude dextran	7.0	91
	Formate dehydrogenase	PEG 1000/potassium phosphate	4.9	94
	Isopropanol dehydrogenase	PEG 1000/potassium phosphate	18.8	98
Saccharomyces carlsbergensis	α-Glucosidase	PEG 4000/dextran T-500	1.5	75
Saccharomyces cerevisiae	α-Glucosidase	PEG 4000/dextran T-500	2.5	86
	Glucose-6-phosphate dehydrogenase	PEG 1000/potassium phosphate	4.1	91
Streptomyces sp.	Glucose isomerase	PEG 1550/potassium phosphate	3.0	86
Klebsiella pneumoniae	Pullulanase	PEG 4000/dextran T-500	2.96	91
	Phosphorylase	PEG 1550/dextran T-500	1.4	85
E. coli	Isoleucyl tRNA synthetase	PEG 6000/potassium phosphate	3.6	93
	Leucyl tRNA synthetase	PEG 6000/potassium phosphate	0.8	75
	Phenylalanyl tRNA synthetase	PEG 6000/potassium phosphate	1.7	86
	Fumarase	PEG 1550/potassium phosphate	3.2	93
	Aspartase	PEG 1550/potassium phosphate	5.7	96
	Penicillin acylase	PEG 4000/crude dextran	1.7	90
Bacillus sp.	Leucine dehydrogenase	PEG 4000/crude dextran	9.5	98
Bacillus species	Glucose dehydrogenase	PEG 4000/crude dextran	3.2	95
Brevibacterium ammoniagenes	Fumarase	PEG 1550/potassium phosphate	0.24	90
Lactobacillus cellobiosus	β-Glucosidase	PEG 1550/potassium phosphate	2.2	98
Lactobacillus sp.	Lactate dehydrogenase	PEG 4000/dextran PL-500	6.3	95

volume ratio of the phases is decreased and thus conditions must be defined to give a sufficiently high partition coefficient for efficient extraction. *Table 4* shows examples of extractive cell debris removal experiments in PEG/dextran systems. In most cases the partition coefficients of these enzymes are between 2 and 10, and the overall yields are 90% or more.

After the first extraction further purification of the protein collected in the top phase is generally accomplished by addition of salts or dextran, thus generating a new phase system. Depending on the requirements for protein purity, or of the following purification procedures, the product can be partitioned either to the new top or bottom phase (32). If further removal of nucleic acids and/or polysaccharides and higher enrichment factors are required the product should be extracted into the top phase and a third separation should be carried out.

For the separation of contaminating proteins no general rules have been defined; conditions are verified experimentally. *Table 5* summarizes an example where the extraction has been coupled to other purification procedures for the recovery of leucine dehydrogenase from *Bacillus sphaericus*. For the large scale purification of β-galactosidase a conventional procedure can be compared with the use of aqueous two-phases. Higgins *et al.* (38) outlined a process of successive

Table 5 Purification of leucine dehydrogenase from *B. sphaericus* (37)

Step	Overall yield (%)	Specific activity (U/mg)	Purification (fold)
1. Cell disruption	100	0.4	–
2. Heat denaturation	100	2.1	5.3
3. PEG/dextran[a] system	97	5.2	2.5
4. PEG/salt system	83	6.5	1.3
5. Diafiltration	70	11.0	1.7
Overall yield:	70%		
Overall purification:	27.5 times		

[a] Crude dextran PL.

centrifugation steps to remove cell debris, precipitated nucleic acids, and precipitated protein product. Veide *et al.* (39) showed that a single aqueous extraction with PEG and salt in which β-galactosidase partitions to the PEG-rich upper phase could be used. Cells, nucleic acids, and most contaminating proteins partition to the salt-rich phase.

After partitioning the polymer can be removed from the protein of interest by ultrafiltration through a membrane of the appropriate type (40). Alternatively, for a PEG-rich phase, salt can be added to establish a new phase system where the protein will partition into the salt phase (26, 41). Other methods include adsorption and subsequent elution, or precipitation with salt (26). Recently, an enzyme extraction method has been devised using a PEG/phosphate system with two sequential extractions which allows recycling of the phase-forming chemicals (see *Figure 14*). The overall product yield was 77% and a 13-fold purification was obtained (42).

7.4 Affinity partitioning

Affinity partitioning, where a biospecific ligand is either present free in the system, or bound to the less polar polymer, offers increased selectivity. Most proteins, nucleic acids, and cell debris will partition to the more polar phase, whilst proteins which bind to the ligand will partition into the less polar phase. Ligands used can be specific for one protein, e.g. substrates, inhibitors, antibodies, and protein A (43, 44), or specific for a group of proteins, e.g. cofactors (45), or triazine dyes (46–48). Some examples are given in *Table 6*; for a more comprehensive list the reader is referred to ref. 26.

Several methods are available for covalent attachment of the ligand to the polymer, usually PEG (28, 56–58). The trichloro-s-triazine method (58) has several disadvantages: it is complex to carry out, uses highly toxic chemicals, and non-specific adsorption to the activated groups may occur. Alternatives previously evaluated include bisepoxy-oxirane activation (55, 56), periodate, and epichlorohydrin (55). After activation, the polymer is incubated with ligand. Unbound ligand is removed by ultrafiltration, gel filtration, or two-phase partitioning. Unreacted groups are then blocked.

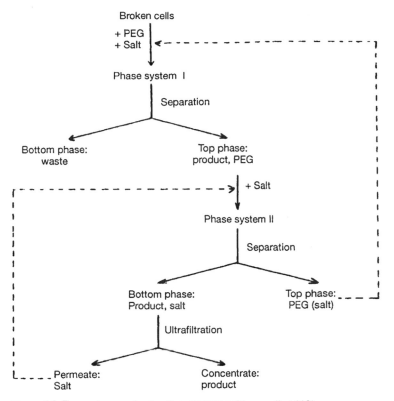

Figure 14 Flow scheme of extractive recovery with recycling (42).

Table 6 Examples of affinity extraction of proteins with PEG/dextran systems

Protein	Ligand	Reference
Trypsin	Trypsin inhibitor	49
Phosphofructokinase	Triazine-dye	50
Formate dehydrogenase	Triazine-dye	51
Glucose-6-phosphate dehydrogenase	Triazine-dye	52
Serum albumin and α-lactalbumin	Fatty acids	53
Albumin and α-fetoprotein	Triazine-dye	54
Thaumatin	Glutathione	55

Factors which must be considered during optimization include the partition coefficient of the protein, the ligand concentration in the phase system, the partition coefficient of the ligand, the dissociation coefficient of the protein-ligand complex, and the number of binding sites that the protein has for the ligand. A potential complicating factor is non-specific binding of contaminants to the ligand-polymer, often due to interactions with the activating group. In affinity partitioning recovery and recycle of the usually costly PEG–ligand complex is a crucial issue. However, the PEG–ligand complex will preferably partition to the PEG-rich phase making the ligand losses in the bottom phase negligible (27).

Acknowledgements

E. L. V. Harris is grateful to J. Fletcher (Millipore), P. Reardon (Romicon), and J. Hickling (Sartorius) for their help in providing information, and in particular to Millipore for providing figures for the ultrafiltration section.

References

1. Freitas, R. D. S. and Cussler, E. L. (1987). *Chem. Eng. Sci.*, **42**, 97.
2. Docksey, S. J. (1986). In *Bioactive microbial products III. Downstream processing*, p. 161. Academic Press, New York and London.
3. Tutunjian, R. S. (1985). *Bio/Technology*, **3**, 615.
4. Anon. (1987). *Bio/Technology*, **5**, 915.
5. Hanisch, W. (1986). In *Membrane separations in biotechnology* (ed. W. C. McGregor), Vol. 1, p. 61. Marcel Dekker, New York and Basel.
6. Blatt, W. F., Dravid, A., Michaels, A. S., and Nelsen, L. (1970). In *Membrane science and technology* (ed. J. E. Flinn), p. 47. Plenum, New York.
7. de Fillipi, R. P. and Goldsmith, R. L. (1970). In *Membrane science and technology* (ed. J. E. Flinn), p. 33. Plenum, New York.
8. Porter, M. C. (1972). *Ind. Eng. Chem. Prod. Res. Dev.*, **11**, 234.
9. McGregor, W. C. (1986). In *Membrane separations in biotechnology* (ed. W. C. McGregor), Vol. 1, p. 1. Marcel Dekker, New York and Basel.
10. Dunnill, P. (1983). *Process Biochem.*, **18**, 9.
11. Narendranathan, T. F. and Dunnill, P. (1982). *Biotechnol. Bioeng.*, **24**, 2103.
12. Pace, G. W., Scherin, M. J., Archer, M. C., and Goldstein, D. J. (1976). *Sep. Sci.*, **11**, 65.
13. Swaminathan, T., Chandhuri, M., and Sirkar, K. K. (1981). *Biotechnol. Bioeng.*, **23**, 1873.
14. Matthiasson, E. (1983). *J. Membr. Sci.*, **16**, 23.
15. Sirkar, K. K. and Prasad, R. (1986). In *Membrane separations in biotechnology* (ed. W. C. McGregor), Vol. 1, p. 37. Marcel Dekker, New York and Basel.
16. Ingham, K. C., Busby, T. F., Sahlestrom, Y., and Castino, F. (1980). In *Ultrafiltration membranes and applications* (ed. A. R. Cooper), p. 141. Plenum, New York.
17. Luong, J. H. T., Nguyen, A. L., and Male, K. B. (1987). *TIBTECH*, **5**, 281.
18. Mandaro, R. M., Roy, S., and Hou, K. C. (1987). *Bio/Technology*, **5**, 928.
19. Brandt, S., Goffe, R. A., Kessler, S. B., O'Connor, J. L., and Zale, S. E. (1988). *Bio/Technology*, **6**, 779.
20. Johns, E. W. (1964). *Biochem. J.*, **92**, 55.
21. Cohn, E. J., Strong, L. L., Hughes, W. L., Mulford, D. L., Ashworth, J. N., and Taylor, H. L. (1946). *J. Am. Chem. Soc.*, **68**, 459.
22. Froud, S. J. (1989). In *Protein purification applications: a practical approach* (ed. E. L. V. Harris and S. Angal). IRL Press, Oxford.
23. Findlay, J. (1989). In *Protein purification applications: a practical approach* (ed. E. L. V. Harris and S. Angal). IRL Press, Oxford.
24. Ingham, K. C. (1984). In *Methods in enzymology* (ed. W. B. Jakoby), Vol. 104, p. 351. Academic Press, New York and London.

Aqueous two-phase partitioning

25. Kula, M. R. (1979). *Appl. Biochem. Bioeng.*, **2**, 71.
26. Albertsson, P. A. (1986). *Partition of cell particles and macromolecules* (3rd edn). Wiley & Sons, New York.

27. Dove, G. B. and Mitra, G. (1986). In *Separation, recovery and purification in biotechnology: recent advances and mathematical modeling* (ed. J. A. Asenjo and J. Hong), p. 93. American Chem. Soc., Washington.

28. Birkenmeier, G., Kopperschlager, G., Albertsson, P. A., Johansson, G., Tjerneld, F., Akerlund, H. E., *et al.* (1987). *J. Biotechnol.*, **5**, 115.

29. Mattiasson, B. and Kaul, R. (1986). In *Separation, recovery and purification in biotechnology: recent advances and mathematical modeling* (ed. J. A. Asenjo and J. Hong), p. 78. American Chem. Soc., Washington.

30. Kula, M. R., Kroner, K. H., and Hustedt, H. (1982). *Adv. Biochem. Eng.*, **24**, 73.

31. Skuse, D. R., Norris-Jones, R., Brooks, D. E., Abdel-Malik, M. M., and Yalpani, M. (1987). Paper presented at the *International Conference on partitioning in aqueous two phase systems*. Oxford, August, 1987.

32. Pfeiffer and Langen (1986). *Aqueous phase systems on the basis of dextran.* Pfeiffer and Langen, Dormagen, FRG.

33. Bamberger, S., Brooks, D. E., Sharp, K. A., Van Alstine, J. M., and Weber, T. J. (1985). In *Partitioning in aqueous two-phase systems* (ed. H. Walter, D. E. Brooks, and D. Fisher), p. 85. Academic Press, New York and London.

34. Hustedt, H., Kroner, K. H., and Kula, M. R. (1985). In *Partitioning in aqueous two-phase systems* (ed. H. Walter, D. E. Brooks, and D. Fisher), p. 529. Academic Press, New York and London.

35. Johansson, G. (1984). *Acta Chem. Scand. Ser. B*, **28**, 873.

36. Kula, M. R., Kroner, K. H., Hustedt, H., Grandja, S., and Stach, W. (1976) German Patent No. 26.39.129; US Patent No. 4, 144, 130.

37. Schütte, H., Kroner, K. H., Hummel, W., and Kula, M. R. (1983). *Ann. N. Y. Acad. Sci.*, **413**, 270.

38. Higgins, J. J., Lewis, D. J., Daly, W. H., Mosqueira, F. G., Dunnill, P., and Lilly, M. D. (1978). *Biotechnol. Bioeng.*, **20**, 159.

39. Veide, A., Smeds, A. L., and Enfors, S. O. (1983). *Biotechnol. Bioeng.*, **25**, 1789.

40. Hustedt, H., Kroner, K. H., Menge, U., and Kula, M. R. (1978). *Abst. 1st. Eur. Congr. Biotechnol.*, Interlaken.

41. Hustedt, H., Kroner, K. H., Stach, W., and Kula, M. R. (1978). *Biotechnol. Bioeng.*, **20**, 1989.

42. Hustedt, H. (1986). *Biotechnol. Lett.*, **8**, 791.

43. Mattiasson, B. (1980). *J. Immunol. Methods*, **35**, 137.

44. Ling, T. G. I. and Mattiasson, B. (1982). *J. Chromatogr.*, **252**, 159.

45. Patton, J. S., Albertsson, P.-A., Erlanson, C., and Borgstrom, B. (1978). *J. Biol. Chem.*, **253**, 4195.

46. Lowe, C. R., Small, D. A. P., and Atkinson, A. (1981). *Int. J. Biochem.*, **13**, 33.

47. Johansson, G., Kopperschlager, G., and Albertsson, P.-A. (1983). *Eur. J. Biochem.*, **3**, 589.

48. Johansson, G. and Joelsson, M. (1985). *Biotechnol. Bioeng.*, **27**, 621.

49. Segard, E., Takerkart, G., Monsigny, M., and Oblin, A. (1973). French Patent No. 7342320.

50. Kopperschlager, G. and Johansson, G. (1982). *Anal. Biochem.*, **124**, 117.

51. Cordes, A. (1985). Dissertation, T. U. Braunschweig, FRG.

52. Kroner, K. H., Cordes, A., Schelper, A., Moor, M., Buckmann, A. F., and Kula, M. R., (1982). In *Affinity chromatography and related techniques* (ed. T. C. J. Gribnau, J. Visser, and R. J. F. Nivard), p. 491. Elsevier, Amsterdam.

53. Johansson, G. and Shanbhag, V. (1984). *J. Chromatogr.*, **284**, 63.

54. Birkenmeier, G., Usbeck, E., and Kopperschlager, G. (1984). *Anal. Biochem.*, **136**, 264.

55. Dunthorne, P. (1988). B.Sc. project in Biotechnology, University of Reading.

56. Head, D. M., Andrews, B. A., and Asenjo, J. A. (1989). *Biotechnol. Tech.*, **3**, 27.

57. Janson, H.-C. (1984). *Trends Biotechnol.*, **2**, 31.

58. Abuchowski, A., Van Es., T., Palczuk, C., and Davis, F. (1977). *J. Biol. Chem.*, **252**, 3578.

Chapter 7
Separation on the basis of chemistry

Lars Hagel

APBiotech, Björkgatan 30, 751 84 Uppsala, Sweden.

1 Introduction

Most chromatographic separations are based on chemical interaction between the solute of interest or impurities to be removed and the separation medium. The exception is separations based upon physical properties such as size (e.g. size exclusion chromatography) or transport in a force field (e.g. electrochromatography).

The chemical interaction may be weak (e.g. employing van der Waals forces) or very strong (e.g. involving formation of chemical bonds as in covalent chromatography). Whenever separation is based upon attractive forces between the solute and the separation medium, we talk about adsorption chromatography (also when the solute is merely retarded). The chemical interaction between the solute and the adsorbent (the chromatography medium) is governed by the surface properties of the solute and the adsorbent and is in most cases mediated by the mobile phase or additives to the mobile phase. Macromolecules such as proteins display a variety of properties and, ideally, a selected set of properties is utilized for obtaining the required selectivity (i.e. relative separation from other solutes) using a separation medium of complementary properties.

This chapter briefly reviews the different types of forces of interaction between solutes and surfaces commonly employed for chromatographic purifications, important properties of solvents, and some basic surface chemical properties of proteins. This, together with a description of some common types of chromatography modes provides a basis for a rational selection of separation mechanism for the purification of proteins and the choice of mobile phase composition to regulate the relative influence of different interaction mechanisms. The separation mechanisms are focused to adsorptive modes with the exception of affinity chromatography which is discussed in Chapter 9.

2 Chemical interaction between surfaces

The different attractive forces acting between molecular and particle surfaces include (1):

- dispersion forces
- electrostatic dipole interactions
- electron donor–acceptor forces
- formation of covalent bonds

All these forces are due to interactions between electric charges (permanent or induced).

Dispersion, or London forces, are caused by induced dipole-induced dipole interactions and are thus classified as a non-specific interaction. This type of non-polar interaction is the dominant force promoting dissolution of non-polar solutes in organic solvents. It may, therefore, also be expected to contribute to non-polar interaction in reversed-phase chromatography (RPC) and hydrophobic interaction chromatography (HIC) (though as we shall see, other interactions may contribute as well).

Electrostatic dipole interactions are caused by orientation of permanent dipoles (Keesom forces) or the interaction between a permanent dipole and an induced dipole (Debye interaction). It can be noted that van der Waals forces include all attractive forces between neutral molecules (e.g. dispersion and electrostatic dipole interactions). Van der Waals forces are short-range inter-actions (typically acting below 25 Å). The strength of the interaction is increased by the polarizability of the molecules and the inverse distance between the molecules, raised to a power of six. It was recently noted that the influence of attractive van der Waals forces may be substantially reduced by bulky substitu-ents on the rather rough surface of chromatography materials (2). The relative interaction strength of a permanent dipole may be estimated from the dipole moment. For large solutes, such as proteins, the interaction must be calculated from the group dipole moment (1).

Electron donor–acceptor interactions span a wide range of interactions, from the simple overlap of electron clouds as in aromatic interaction, the sharing of lone electrons as in hydrogen bonds, to the overlap of orbitals to form ionic bonds. Aromatic adsorption effects are believed to be due to interaction be-tween the electrons in the π-orbitals (e.g. as in the aromatic ring) or between electrons in π-orbitals and electron-rich parts of other molecules. Aromatic interaction between a solute and a ligand may be quenched by addition of π-electron-rich modifiers, such as pyridine. Hydrogen bonds are formed when a proton is shared between two electronegative atoms, e.g. O, N, or F. Hydrogen bonds are present in all aqueous systems. Hydrogen bonding between a solute and an adsorbent is therefore carried out in non-aqueous solvents (e.g. metha-nol). The influence of hydrogen bonding is increased by so called structural enhancers (e.g. potassium sulfate) and decreased by structural breakers (e.g. urea, formamide). Electron-rich parts of proteins (e.g. histidine residues) may form a coordination complex with immobilized metal ions. Pure ionic forces, attractive or repulsive, are present for charged surfaces. This interaction is a long-range interaction, i.e. typically extending up to 100 Å in solutions of low ionic strength

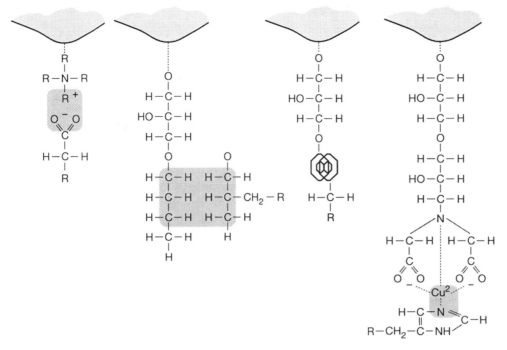

Figure 1 Some molecular interactions employed for chromatography of proteins. The figure illustrates, from left to right; ionic interaction (e.g. between a quaternary ammonium ligand and an aspartic acid residue), hydrophobic interaction (e.g. between a butyl ligand and a leucine residue), aromatic interaction (e.g. between a phenyl ligand and a phenylalanine residue), and formation of a coordination complex (e.g. between an immobilized copper ion and a histidine residue).

(high ionic strength will screen ionic interactions). Ion–ion forces are rather strong (e.g. two orders of magnitude higher than dipole–dipole interaction) and furthermore, do not decrease as rapidly with distance as the latter force. The dielectric constant of the solvent as well as concentration and type of charged solutes will influence the interaction.

Formation of covalent bonds may be classified as one extreme of electron donor–acceptor interaction. However, since this interaction involves a transition from intermolecular to intramolecular forces it is treated separately in this chapter. One example is given by the intramolecular formation of disulfide bonds in proteins. The formation and breakage of this type of bond is affected by the redox properties of the solution and the presence of SH-containing solutes. Sometimes, it is necessary to add special protective groups to prevent formation of undesired covalent bonds.

Some types of chemical interactions used for chromatography are illustrated in *Figure 1.*

Adsorptive separations are based on the selective manipulation of some of these interactions, while reducing the influence of other effects, through the use of purposely made adsorbents and carefully selected chemical conditions.

The interaction is driven by thermodynamics, i.e. to minimize the total free standard energy of the system, $\Delta G°$, as given by enthalpy effects, $\Delta H°$, and entropy effects, $\Delta S°$, at a certain absolute temperature, T:

$$\Delta G° = \Delta H° - T\Delta S° \tag{1}$$

The degree of retention in adsorptive chromatography may be related to the change in standard free energy for the transfer of solutes from the liquid phase to the adsorbent phase, by:

$$K_D = \exp(-\Delta G°/RT) \tag{2}$$

where K_D is the distribution coefficient and R is the gas constant. Retention will thus be achieved for systems where the enthalpy is decreasing and/or systems where the entropy is increasing to give a net decrease in free energy. As discussed below, different behaviour has been noticed for various modes of chromatography, e.g. ion exchange chromatography (IEC) and HIC. The free energy change involved in a reversible adsorption process is in the order of -15 kJ/mol corresponding to a distribution coefficient of 450.

The type and degree of interaction will depend upon properties of the solute, the mobile phase, and the chromatography surface. Theoretically, it would be possible (from careful characterization of the solvent properties, the adsorbent properties, and the chemical structure of the molecule) to identify and quantify the interactions, calculate the total free energy change upon interaction at most preferential orientation, and obtain an estimate of the distribution coefficient from Equation 2. Even for small solutes interacting with a surface, the many types of interactions, and the difficulty in quantifying these makes it difficult to obtain reliable estimates of retention this way (1). Instead, qualitative predictions of retention in chromatography have been made from studying the influence of different substituents in homologous series of solutes having small to medium size (e.g. organic solutes and peptides up to 50 amino acid residues) (1, 3). Still, for large molecules such as proteins, the heterogeneous properties of the surface, displaying polar, non-polar, aromatic, and charged patches, together with the three-dimensional structure which adds an orientation effect to the interaction, makes such predictions very unreliable. Also, the interaction between surfaces may induce properties not normally displayed or cause conformational changes (or even disrupt the structure of the molecule) which further complicates the situation.

Nevertheless, we may from global or lumped information about the surface properties (i.e. such as isoelectric point, hydrophobicity index, relative polarity, etc.) of different molecules make qualitative predictions about how the molecules are likely to be affected under various, carefully selected, conditions and how this will influence their separation relative to each other by different chromatographic techniques.

3 Basic properties of proteins

Proteins are composed of amino acids having non-polar, polar, aromatic, and charged residues (e.g. see ref. 4 for a thorough description of properties of proteins). Proteins are compact structures with only a fraction of the amino acids exposed on the surface in their native form. The properties of the exposed amino acids are a function of their natural environment; for aqueous proteins charged and polar residues are common, whereas for membrane proteins non-polar (lipophilic or hydrophobic) amino acids are normally found on the surface. However, it must be stressed that most proteins expose many different types of amino acid residues on the surface. For instance, about 55% of the total accessible surface area of globular proteins in their native state is non-polar (5). The other side of this coin is that non-polar amino acids (hydrophobic) may be found in the interior of hydrophilic proteins while the opposite may be true for hydrophobic proteins. Thus, a disruption of protein structure (i.e. as sometimes found for RPC) may expose a new set of surface properties.

The charged amino acids may either be basic or acidic. This results in the property of proteins being amphoteric; at low pH proteins are positively charged due to the basic (e.g. amino) groups and at high pH proteins are negatively charged due to the acidic (e.g. carboxylic) groups. The pH where the net charge is zero is called the isoelectric point. It is important to note that at this point both negatively and positively charges are present on the surface and furthermore negatively charged groups and positively charged groups may be found on either side of the isoelectric point. The change of charge of a protein as a function of pH is called its titration curve and may be determined by electrophoresis (i.e. see ref. 6).

Charges on the surface of proteins may thus be affected with the aid of pH. Another way is to let the protein molecule interact with a charged solute, e.g. (sodium) dodecyl sulfate which is negatively charged or tetrabutylammonium (sulfate) which is positively charged, to form an ion pair. This will reduce the polar nature of charged proteins.

Since the surface of a protein is heterogeneous, macroscopic properties will not be adequate for detailed explanations or predictions of interactions. The local interaction distances in IEC are of the order 7–16 Å (7, 8) which results in only a few of the charged sites being involved in the interaction (e.g. typically three to six charges, called the characteristic charge). Also, the distribution of hydrophobic patches on the protein as related to the density of hydrophobic ligands on the chromatographic surface will have a decisive influence on the adsorptive properties. From molecular modelling the local surface properties of proteins may be estimated and used to get qualitative and quantitative information about potential interactions with chromatography media as recently shown for HIC (9).

The stabilization of protein (three-dimensional) structure is a result of competition between solvation and intraprotein forces. The latter is believed to be composed of van der Waals forces and hydrogen bonding (10). It was noted that hydration of non-polar residues is energetically favourable and may lead to denaturation (e.g. as noticed for cold denaturation) (10).

4 Properties of the solvent

The solvent used in chromatographic separations needs to fulfil the following basic requirements:

(a) It should be a good solvent for the solute (i.e. to avoid precipitation or aggregation).

(b) It should provide complementary properties to the chromatographic sorbent (i.e. to allow for a separation the solvent and sorbent must have different properties).

(c) It should be compatible with the detection system used (e.g. UV transparent).

For hydrophilic proteins a polar solvent must be used. The tolerance to organic solvents varies between proteins and this can be tested by *Protocol 1*. In order to assure stability of the protein in solution the pH and the ionic strength may need to be within certain intervals (i.e. often a pH close to the isoelectric point is avoided since this may promote precipitation) and cofactors may need to be present. These initial conditions may set limits for the selection of suitable techniques (e.g. use of complexing agents). Water is a strong hydrogen bond forming solvent that will prevent the use of water as solvent for separations using a hydrogen bonding sorbent. In this case a solvent such as methanol may be used. Properties of the solvent may be regulated by the addition of co-solvents or other additives. The properties of some solvents used in chromatography are listed in *Table 1*. The table illustrates that water is a good solvent for polar substances as opposed to propanol, while propanol is more suitable for non-polar solutes as seen by the higher value of the dispersion coefficient. Acetonitrile has a poor hydrogen bonding ability as compared to the other solvents. Tetrahydrofuran has properties that deviate too much from water and is not miscible with pure water. The viscosity of propanol will create high back pressures in column chromatography.

Additives may be added to the solvent for a variety of purposes (for example in order to increase the solubility of proteins, to reduce one type of undesired

Table 1 Hildebrand solubility parameters for some solvents (1)

Solvent	Polarity[a]	Dispersion	Dipole	Proton-donor	Proton-acceptor	Viscosity[b] (cP)
Tetrahydrofuran	9.9	7.6	4	0	3	0.55
Propanol	10.2	7.2	2.5	4	4	2.30
Ethanol	11.2	6.8	4.0	5	5	1.20
Acetonitrile	11.8	6.5	8	0	2.5	0.37
Acetic acid	12.4	7.0				1.26
Methanol	12.9	6.2	5	5	5	0.60
Water	21	6.3	Large	Large	Large	1.00

[a] Polarity refers to effects from permanent dipoles.

[b] Viscosity data is taken from ref. 11.

interaction with the adsorbent, or to elute the protein from the adsorbent). Different types of additives used for elution have different elution strengths. The elution strength may be specific, as for affinity chromatography, or solely related to the ionic strength, as for IEC. Examples of additives that have been found useful for different chromatographic techniques are found under the optimization heading of the technique in question.

The effect of water and water structure on association of molecules, especially in hydrophobic types of interaction, have been under much debate and still the hydration forces of water are not well understood (e.g. see refs 2, 10, 12). The present interpretation is that association of molecules has more to do with chemical and physical nature of surfaces than with water structure (2).

5 Properties of chromatography media

Chromatography media used for separations based upon chemistry may be termed adsorptive media, or adsorbents, since the process involves an active adsorption step, as opposed to gel filtration where the solutes are separated by their size-dependent permeation of a porous medium. However, in order to create a large surface area for sorption (and thus a high capacity) most adsorptive chromatography media for preparative use are porous and size exclusion will be a mechanism that is superimposed on the adsorptive retention. The adsorptive process may generally be described by the equation:

$$P + C-L- \leftrightarrow P-L- + C \qquad [3]$$

where P is the protein, P-L is the complex between the protein and the ligand, L, attached to the chromatographic surface and C is the leaving group. The forward reaction is characterized by an association constant, k_{ass}, and the reverse reaction by a dissociation constant k_{diss} ($k_{ass} = 1/k_{diss}$). An affinity interaction having an association constant larger than 10^5 M^{-1} is classified as a strong interaction (the dissociation constant is smaller than 10 μM, and the affinity is said to be in the micromolar range). This corresponds to a distribution coefficient of 100 and a change in Gibbs free energy of −11.3 kJ/mol for low loadings of a solute of molecular mass 50 000 adsorbed to a sorbent having a maximum capacity of 50 g/ml.

The properties of different adsorbents are generally designed to facilitate one selected mechanism. This mirrors the properties of the solutes and may employ strong Coulombic forces as in ion exchange chromatography, dispersion forces as in HIC and RPC, charge transfer as in hydrogen bonding, formation of a coordination complex as in immobilized metal affinity chromatography, or formation of covalent bonds as in covalent chromatography (or a combination of forces as in affinity chromatography). These are the most widely used separation techniques for proteins but still only some of the techniques that may be employed (e.g. see ref. 13 for a basic review of techniques for protein chromatography).

Since it is virtually impossible to create an totally inert base matrix also this will have some properties that will contribute to the separation and the result will sometimes be significantly affected by unexpected contributions from the matrix. These effects are sometimes referred to as non-specific (instead of non-expected). However, unless the effect is caused by induced dipole–dipole interactions as described above the interaction is in principle specific!

The retention in adsorptive chromatography is expressed by the retention factor, k', which corresponds to the amount of solute in the stationary phase over the amount of solute in the mobile phase. The retention factor is related to the distribution coefficient by:

$$k' = K_D \cdot \Phi \qquad [4]$$

where Φ is the phase ratio (i.e. the volume of the stationary phase, V_S, over the volume of the mobile phase, V_M). Finally, the retention volume, V_R, of a particular solute is given by the retention factor as:

$$V_R = V_M \cdot (k' + 1) \qquad [5]$$

Thus, the retention factor varies with the distribution coefficient which in turn is regulated by the thermodynamics of the reversible sorptive process. The retention is affected by the mobile phase composition and a general equation is (7):

$$\log k' = A + B \log m_s + C m_s \qquad [6]$$

where A is a system constant, B is the electrostatic interaction parameter (being proportional to protein characteristic charge), and C is the hydrophobic interaction parameter (this will vary with hydrophobic contact area). The different influence of the concentration of mobile phase modifier, m_s (e.g. salt in ion exchange chromatography $[C = 0]$, and HIC $[B = 0]$, and organic modifier in RPC, $[B = 0]$) on the retention for different chromatographic modes should be noted. The retention in adsorptive chromatography is influenced by the number of interaction points and hence the size of the molecule. Therefore, different behaviour may be expected for small solutes, where a one-to-one relationship between interaction points on the surface of the solute and the surface of the adsorbent may be expected and for large solutes where a number of interaction points are engaged in the adsorption process. Thus, for proteins possessing multiple sorption sites high concentrations of the mobile phase are required to assure that all binding sites are, at the same time, successfully competed for by the desorbing molecule to release the protein. At this concentration of desorber the equilibrium for sorption is heavily shifted to the left (Equation 3) and no re-adsorption takes place (this is sometimes referred to as the on–off effect). The composition of the mobile phase needs to be changed quite drastically to desorb proteins differing in characteristic charge and hence, gradient or step elution are common for separation of protein mixtures.

Separation by an adsorptive process may be divided into five phases as illustrated by *Figure 2*.

(a) **Equilibration of the adsorbent**. This is done using a mobile phase which will facilitate adsorption of the protein of interest (e.g. pH, counter ion, ionic strength).

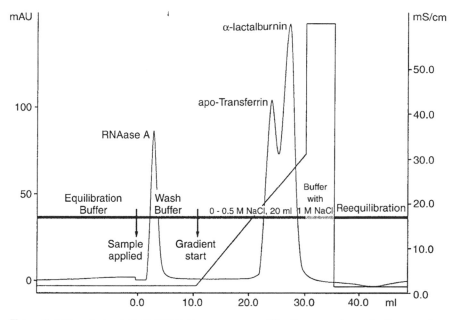

Figure 2 Different steps in an adsorptive process, equilibration, sample application, wash, desorption, wash and re-equilibration. Intermittent CIP (cleaning in place), is also carried out, depending upon the degree of fouling. The example shows separation of a protein mixture on Q Sepharose Fast Flow. Reproduced from ref. 14. Copyright 1997 Pharmacia Biotech AB.

(b) **Sample application**. This is done in a mobile phase that will solubilize the protein and also facilitate quantitative adsorption, promoting a high association constant (see *Protocol 1*).

(c) **Wash**. This is done using a mobile phase which will wash away non-adsorbed or weakly adsorbed proteins but not affect the protein of interest. An erroneous choice of wash solution may result in that the target protein starts to elute, with an impaired resolution or unpredicted separation as result.

(d) **Desorption of the protein**. This is done using a mobile phase that will increase the dissociation constant so that the protein is desorbed. This may be done in three different modes. A step change or a gradual change of the composition of the mobile phase if a large change in the dissociation constant is required (as often is the case with proteins). If the adsorption step is carried out in a mobile phase where the association constant is low (e.g. 10^4 M^{-1}) the protein sample may be eluted by an isocratic elution step. In this case the concentration of the mobile phase is kept constant. Solutes are eluted as a result of the competition between protein and components of the mobile phase.

(e) **Wash**. After the protein is eluted the more strongly retained components are desorbed by changing the pH, ionic strength, or concentration of organic solvents.

A pH 5.0 pH 5.5 pH 6.0 pH 6.5 pH 7.0 pH 7.5 pH 8.0 pH 8.5 pH 9.0 pH 9.5

B 0.05 M 0.1 M 0.15 M 0.2 M 0.25 M 0.3 M 0.35 M 0.4 M 0.45 M 0.5 M

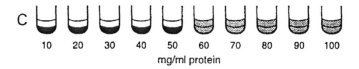

C 10 20 30 40 50 60 70 80 90 100

mg/ml protein

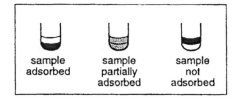

sample adsorbed   sample partially adsorbed   sample not adsorbed

Figure 3 Selecting initial conditions for binding and elution in adsorptive chromatography. The figure illustrates the principle of selecting the initial conditions for binding and desorption of an anion exchanger. The first step is to determine a suitable pH where the protein is negatively charged and binds to the adsorbent (A). In this case, binding pH ≥ 7.5, elution pH ≤ 5.0. The second step is to determine the salt concentration of the buffer where the protein is adsorbed and desorbed (B). In this case, binding at %$B \leq 0.15$ M and elution at %$B \geq 0.3$ M. The third step is to determine the static capacity (i.e. maximum binding capacity) of the adsorbent at binding conditions (C). In this case, the maximum capacity is 50 mg protein/ml adsorbent. Adapted from refs 6 and 15 with permission.

(f) **Cleaning in place** (CIP). If proteins or other sample components are strongly adsorbed to the adsorbent or column components the column needs to be cleaned. This is often done by extreme pH (e.g. 0.5 M sodium hydroxide, acid, 0.1 M HCl, or other additives). These conditions are often harsh and may set a limit for the lifetime of the adsorbent.

(g) **Regeneration of the adsorbent**. The adsorbent needs to be transferred to the reference state again before a new separation is carried out.

The experimental conditions to be used for adsorption and desorption are related to the separation principle employed. However, a general methodology may be applied as described in *Protocol 1* and illustrated in *Figure 3*.

Protocol 1

Screening of experimental conditions

This protocol may be followed to screen suitable experimental conditions for adsorption and desorption conditions in all types of adsorptive chromatography steps (see *Figure 3*).

Equipment and reagents

- Protein solution
- Adsorbent (e.g. ion exchanger, HIC medium, etc.)
- Adsorption buffers of various composition (e.g. pH, ionic strength, organic modifier)
- Desorption buffers of various composition (e.g. pH, ionic strength, organic modifier)
- Assay of protein concentration and protein activity

Method

1   Dissolve the protein in each of the adsorption buffers to a known final concentration of 2–10 mg/ml. Alternatively, add a defined (small) volume of a protein stock solution to each of the adsorption buffers. The dilution of adsorption buffers should not exceed 1:1. Visually inspect the solutions for precipitates—only conditions providing clear solutions should be used.[a] Determine the amount of active target protein in the solution under the conditions (i.e. incubation time and temperature), as planned for the purification step. Reject conditions for which the protein is not stable.

2   Prepare the adsorbent according to the manufacturer's instruction and finally wash the adsorbent with three portions (one portion is equal to one gel volume) of starting buffer by suction on a glass filter:

   (a) Transfer a defined amount of drained gel (e.g. 2 g) into a beaker.

   (b) Add a defined amount of protein solution (e.g. 10 ml) to the beaker.

   (c) Mix gently for 10 min and let the gel settle.

   (d) Measure the amount of active protein in the supernatant.

   (e) Plot the apparent[b] amount of adsorbed protein per gram of gel from (conc. in solution − conc. in supernatant) × volume of solution/amount of gel versus the buffer composition.

3   Repeat steps 1 and 2 with the protein dissolved in desorption buffers. In case several parameters are studied they may be varied one at a time as illustrated in Figure 3.

4   From the plot obtained in step 2e, determine the conditions (e.g. pH, ionic strength, % organic modifier) necessary for complete adsorption of the target protein (or alternatively complete adsorption of contaminating solutes only). Determine the conditions (e.g. ionic strength and pH for IEC) needed for desorption of the protein in active form from the plot obtained in step 3.

5   The static capacity of the gel may be estimated by varying the protein concentration and keeping other conditions constant.[c] This will in principle yield the adsorption isotherm, i.e. the amount of adsorbed protein as a function of initial concentration at constant temperature and mobile phase composition.

Alternative procedures

The procedure above may readily be adapted to small pre-packed columns (e.g. HiTrap®) or performed using laboratory packed columns (e.g. using PD-10). In this case a more realistic picture of the dynamic adsorptive properties of the adsorbent is obtained.

[a] Precipitation will increase the turbidity of the solution due to light scattering and this may be detected as a decreased transmission, or an apparent increase in absorption, using a photometer at 360 nm.

[b] In order to calculate the exact amount of protein adsorbed the interstitial and interparticle liquid volume of the drained gel need to be known since this liquid dilutes the protein solution. This may be estimated by calculating the dilution of a non-adsorbing solute (e.g. a protein at high ionic strength for an IEC step).

[c] The amount of adsorbed protein calculated in this way may be used to give an indication of the different capacities of various media under static conditions. However, it must be realized that this figure may be an overestimate of protein adsorption capacities under dynamic conditions (e.g. column chromatography) by a factor of 2–6 and furthermore, this evaluation does not account for differences in rates of mass transfer and adsorption/desorption kinetics between different media.

Some chromatographic techniques that have gained wide popularity for adsorptive purification of proteins are IEC, HIC, RPC, and immobilized metal affinity chromatography (IMAC). Some guidelines to experimental work with these and some other interesting purification principles of complementary properties, such as covalent chromatography and hydroxyapatite chromatography, will now be given.

5.1 Ion exchange chromatography

IEC is one of the most widely used techniques for separation of proteins on the basis of Coulombic interaction. This is due, in part, to the fact that many physiologically important proteins have a large amount of charged amino acid residues on their surface. In addition, the variation of surface charge of proteins with change in pH provides a simple means of regulating the adsorptive properties of the solute. Finally, the desorption of proteins (in an active form) is mostly straightforward by increasing the ionic strength of the eluent (6).

Two different modes of ion exchange exist. One is cation exchange, CIEC, where the adsorbent is negatively charged (e.g. due to the content of sulfonic acid ligands). This is used for separating positively charged proteins, usually predominantly at a pH below their isoelectric point, and is useful for many serum proteins. The other mode is anion exchange, AIEC, where the adsorbent is positively charged (e.g. due to the presence of amino groups). If the charge of the ion exchanger varies with pH within the operating range (e.g. pH 4–10) the ion exchanger is said to be a weak ion exchanger.

5.1.1 Separation principle

The adsorption of charged macromolecules such as proteins to an oppositely charged adsorbent has been explained from a model where the counter ions (denoted C in Equation 3) on the adsorbent are displaced by the interacting charged patches of the macromolecules. The combined effect of Coulombic and van der Waals interactions has been used to explain retention of proteins in IEC (8). The (statistical) number of interaction points depends on the conditions but typical figures reported are in the range of 3–6. The adsorption of large solutes may also result in shielding of charged groups. From theoretical relationships, protein separations by IEC may successfully be modelled in both analytical and preparative scale (16, 17).

Though ionic interaction is the macroscopic explanation for ion exchange processes the mechanism seen on a molecular level requires some considerations. It was recently demonstrated that adsorption of proteins to some ion exchangers can be an endothermic process and results in a net gain in enthalpy of up to 40 kJ/mol (18). Thus, in order for the adsorption to take place the increase in entropy must more than balance this gain in enthalpy (increased retention with increased temperature was noted). The explanation given is that counter ions bound to the solute and the adsorbent or waters of hydration are released upon adsorption and this causes an increase in entropy. The increase is larger for solutions of low ionic strength which promotes a stronger adsorption than with solvents of high ionic strength.

5.1.2 Critical factors

Critical factors in IEC include selection of pH to, if possible, introduce as large charge difference as possible between the target solute and the contaminants. This will provide the highest selectivity and will reduce the constraints on the other parameters. Another important factor is the pore dimensions of the matrix (i.e. a difference in size between the solute and contaminants may be exploited by selecting a sorbent of suitable pore size). If large amounts of solute are to be purified the pore dimension must be at least four times the size of the molecule in order to allow access to the internal surface area of a porous matrix which may be 100 times larger than the external surface area. The radius of a protein having a molecular mass of 50 000 is approximately 30 Å which implies that the pore radius of the adsorbent should exceed 120 Å to allow for significant permeation. For larger proteins an adsorbent having a pore radius of 300 Å is more suitable. Non-porous media may be employed in analytical applications where low amounts of protein are applied. Other factors that will influence the separation are, in decreasing importance, gradient slope (i.e. the change of mobile phase composition, $\%B$, per unit bed volume), the load (i.e. applied amount relative to the capacity of the adsorbent), particle size, column length, and fluid velocity. The salt concentration as such, or rather the elution strength of the eluent, is an important factor as seen from Equation 6. Sometimes specific elution properties of different salts have been reported. However, some of these effects have later been explained by their stoichiometric influence on ionic strength (19). Therefore simple salts, such as sodium chloride, may generally be used in a first attempt to optimise separation of proteins in IEC.

5.1.3 Optimization of binding and elution conditions

Before an ion exchange experiment is designed, some basic knowledge about the sample needs to be collected. This includes the activity of the protein at different pH and ionic strengths, and whether additives, such as cofactors are necessary. The charge of the target molecule and the contaminants at various pH will help in deciding what mode to run (i.e. AIEC or CIEC). Finally the size of the molecule and contaminants will give information about optimal pore size of the ion exchanger (i.e. large contaminants may be excluded from the porous matrix to facilitate purification and to prevent contaminants from consuming valuable capacity).

If data are not available, suitable experimental conditions may be found by screening as described in Protocol 1, provided the target protein can be selectively assayed. The following conditions may be used as a starting point in this protocol:

(a) **Adsorption buffer**. 3, 2, 1.5, and 1 pH unit below the isoelectric point, for cation exchange or above the isoelectric point, for anion exchange. If the isoelectric point is unknown pH 2, 4, 6, 8, and 10 may be selected as a first start.

(b) **Desorption buffer**. 50, 100, 200 mM NaCl in adsorption buffer (NB: check pH after addition of salt!).

(c) **Buffers for anion exchange** may be prepared from piperazine (pH 5.0–6.0), bis-Tris propane (6.4–7.3), Tris (7.6–8.5), diethanolamine (8.4–8.8), and ethanolamine (9.0–9.5).

(d) **Buffers for cation exchange** may be prepared from maleic acid (pH 1.5–2.5), citric acid (2.6–3.6), formic acid (3.8–4.3), acetic acid (4.8–5.2), malonic acid (5.0–6.0), and phosphate (6.7–7.6). All data for buffers are taken from ref. 15.

5.1.4 Experimental set-up and optimization

When the basic conditions for adsorption and desorption conditions have been established, a preliminary experimental set-up can be made (e.g. particle size, column dimensions, etc.). Suggested starting conditions as given in *Table 2* may be useful in this process. Typical starting conditions for AIEC are an aqueous buffer of a concentration below 50 mM having a pH of 1 unit above the isoelectric point of the protein. Starting conditions for a cation exchanger are of similar concentration but with a pH of 1 unit below the isoelectric point of the protein. The elution conditions are typically a linear gradient from the starting conditions to a final buffer equal to the starting buffer but containing 1 M NaCl. Gradient length is typically equal to 10–20 column volumes. The initial con-

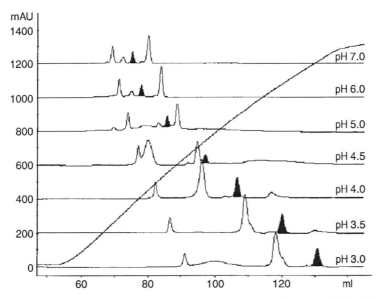

Figure 4 Method scouting of suitable pH for separation of proteins by IEC. Sample: protein mixture of α-chymotrypsinogen A, cytochrome *c* (shaded), lysozyme, transferrin, ovalbumin, and β-lactoglobulin. Buffer: automatically prepared from buffer mixture composed of 0.03 M Na_2HPO_4, 0.03 M NaFormate, and 0.06 M NaAcetate titrated with 0.1 M HCl. Column: Mono S HR 10/10. System: ÄKTA™ explorer. Reproduced from ref. 21. Copyright 1996 Pharmacia Biotech AB.

ditions may require further optimizations to give desired purity and yield and this may be performed according to *Protocol 2*. The optimization is carried out using a chromatographic system, preferably a fully automated system with buffer blending and method scouting properties (21). An example of a typical result for scouting of suitable pH for a cation exchange purification is shown in *Figure 4*.

Table 2 Starting conditions for experimental optimization of large scale preparative ion exchange chromatography (20)

Condition	Capture[a]	Intermediate[a]	Polishing[a]
Particle size (μm)	> 100	30–100	15–50
Bed length (cm)	10	15	15
Velocity (cm/h)	> 300	> 200	> 100
Residence time (min)[b]	> 5	> 5	> 5
Gradient, 0–1 M NaCl in	10 CV[c]	20 CV	20 CV
Sample load (% of max)	< 80	< 50	< 30

[a] Capture, intermediate purification, and polishing is explained in Section 5.6.

[b] Residence time (i.e. time for a non-retarded solute band to travel through column) may require a low velocity during sample application.

[c] CV = column volume, or bed volume.

Protocol 2

Optimization of experimental parameters in adsorptive chromatography

The optimization of parameters in different modes of adsorptive chromatography are basically performed in a similar fashion, irrespective of technique.

Equipment and reagents

- Chromatography system
- Packed column: the size of the column is dependent upon the application but column lengths for IEC and HIC are typically 5–15 cm long and for RPC 10–20 cm long. The diameter is selected to give the total amount of gel required according to the capacity needed for the process.
- Two pumps capable of delivering the required flow rate at the back pressure generated by the system. (One pump system where the gradient is formed by a switching between the two mobile phases generally has lower performance than a two-pump system forming the gradient on the high-pressure end. For low-pressure systems a simple gradient mixer and one pump may be sufficient.)
- Detector (e.g. UV detector at 280 nm to detect proteins, a conductivity detector to trace the gradient)
- Controller (to control the formation of the gradient, automatic sample application, switching of valves, and control of fraction collection)
- Injection device (may be a loop injector for application of small volumes or a pump for application of large volumes)
- Fraction collector

- Recorder (if the controller is PC based it may be able to sample the signal such that a recorder is then optional)
- Equilibration buffer (e.g. to transfer the adsorbent to correct pH and ionic strength): this is often identical to the binding buffer
- Binding buffer (conditioned to promote binding/non-binding, for example by buffering to assure correct pH and having a suitable ionic strength): see *Protocol 1*
- Sample solution in binding buffer
- Wash buffer, buffer A (selected to desorb impurities, not the target protein): often identical to binding buffer
- Elution buffer, buffer B (selected to desorb target protein but not inactivate the protein or cause aggregation, precipitation, etc.)
- Cleaning solution (e.g. a high ionic strength buffer for IEC): sometimes harsh conditions are needed to clean the column, e.g. by pepsin or 0.1–0.5 M NaOH
- Regeneration solution (to restore the column to the reference state): this solution is often identical to equilibration buffer, but of higher concentration (to speed up the regeneration process)

A. Stepwise optimization

The first steps of the optimization are to select binding conditions and choose the type of matrix and ligand. This is done in the screening of experimental conditions (i.e. according to *Protocol 1*). The major optimization step is then to select the gradient slope (i.e. step 4) to give the required resolution. An excessive resolution may be exploited to increase the sample load or flow rate or decrease the column length or gradient length to increase the productivity (i.e. the amount of purified product per unit time and volume of adsorbent).

1 Select binding conditions (e.g. pH and ionic strength). The single largest influence on the selectivity for a specific adsorbent is given by the composition of the binding

Protocol 2 continued

buffer. This is most often selected to preferentially favour interaction between the target protein and the adsorbent. It is important to assure that the association constant is high, otherwise the solute may only be retarded and partially eluted by the wash buffer (this will cause non-reproducible retention times).

2   Select the type of matrix. This is often done in concert with step 1. However, properties of the sample may limit selection of binding conditions (e.g. solubility constraints) and therefore binding conditions are often selected first. Adsorbents of high capacity will allow application of larger amounts of protein before overload effects are noted (i.e. up to 30% of the maximum capacity). Small particle size gives high resolution (due to low zone broadening) at the expense of cost and applicable flow rate. Particle size of 10–30 μm is mostly sufficient for laboratory work while 15–150 μm is suitable for large scale purifications (see *Table 2*). Resolution may conveniently be regulated by the gradient (see below).

3   Choose column length, typically 5–15 cm for IEC and HIC, 10–20 cm for RPC of proteins. For extreme situations longer columns may be needed.

4   Select the gradient slope.[a] This parameter is the by far most important parameter of the running conditions. A more shallow gradient will result in larger spacing between the protein zones but also a widening of the zones (i.e. dilution). The net effect is that gradient volumes of between 10 and 20 column volumes are optimal for elution. The actual slope depends upon the start and final composition (i.e. to prevent early desorption of target molecule and to allow quantitative desorption of the target molecule only) as tentatively obtained in the binding study (see *Protocol 1*). Start by running a wide gradient, i.e. 0–100% B. Then fine-tune the gradient to ideally elute only the target protein (and closely related impurities) in the gradient. Impurities are thus either unbound or not desorbed until the second wash step. Check the optimal conditions by running a sample dissolved in the optimal equilibration buffer.

5   The flow rate has a secondary influence on the resolution but must be low enough to allow sufficient time for the solutes to interact with the adsorbent (residence time).

B. Optimization by scouting

Screening of initial experimental conditions may be performed on a pre-packed column according to the scheme above using the conditions described in *Protocol 1*. This is called scouting and many modern chromatography instruments have built in features comprising this tool.

C. Robustness

After the conditions have been selected, the robustness of the purification may be tested by determination of the influence (i.e. a variation of ± 10%) of different critical experimental parameters using a reduced factorial design (20). Some of the parameters that may be tested include, mobile phase composition (e.g. pH, ionic strength, % of organic modifier), wash volume, sample load, and gradient length.

[a] Please note that retention will normally vary with ligand density (e.g. ionic capacity). Thus, the concentration of elution buffer may need to be adjusted accordingly to give comparable results in terms of retention in number of column volumes.

5.1.5 Chromatofocusing

Chromatofocusing is a special type of IEC where the pH of the eluent is continuously changing. This is a result of the fact that a buffer mixture, having a range of acid constants, pK_a, will titrate a weak ion exchanger during the passage through the bed. The local pH of the mobile phase will change and thus a pH gradient is created. In the most widely used mode the pH gradient is decreasing and the adsorbent is an anion exchanger. The pH of the equilibration buffer is basic, most proteins are negatively charged and therefore adsorbed to the positively charged anion exchanger. The pH of the elution buffer is made acidic and starts to titrate the ion exchanger. The buffer will as a result of the titration attain a more basic pH that will gradually decrease as more buffer is supplied to the column. When the pH of the mobile phase is lower than the isoelectric point of the protein the protein start to move with the fluid. In this way the proteins are eluted according to decreasing isoelectric point. The protein bands are self-sharpening since the pH gradient moves more slowly than the mobile phase and remarkable separations (i.e. of proteins differing by only 0.1 pH unit in isoelectric point) may be achieved using a shallow gradient (22). To achieve this high resolvability, a complicated buffer mixture of ampholytes (e.g. PolyBuffer) is needed. However, for large scale purifications simpler systems are required. As recently demonstrated, large scale purification of recombinant human growth hormone from *E. coli* was successfully performed by chromatofocusing on DEAE Sepharose Fast Flow using a simple buffer mixture of piperazine and triethanolamine (23).

5.2 Hydrophobic interaction chromatography

Since a high proportion of proteins, even of hydrophilic nature, have a substantial amount of hydrophobic patches on the surface, HIC may be used for purification of a large variety of proteins. Adsorbents for HIC generally have alkyl- or aryl-ligands (e.g. butyl, octyl, or phenyl groups) attached to the surface (24). The ligand density is low as compared to adsorbents for RPC (e.g. in the

Table 3 Starting conditions for experimental optimization of large scale preparative hydrophobic interaction chromatography using a phenyl adsorbent (20)

Condition	Capture[a]	Intermediate[a]	Polishing[a]
Particle size (μm)	> 100	30–100	15–50
Bed length (cm)	10	15	15
Velocity (cm/h)	> 300	> 200	> 100
Residence time (min)[b]	> 5	> 5	> 5
Gradient, 1–0 M NH$_4$Cl in	10 CV[c]	20 CV	20 CV
Sample load (% of max)	< 80	< 50	< 30

[a] Capture, intermediate purification, and polishing is explained in Section 5.6.

[b] Residence time (i.e. time for a non-retarded solute band to travel through column) may require a low velocity during sample application.

[c] CV = column volume, or bed volume.

range of 5–50 μmol/ml gel) which is advantageous for preserving protein structure. Since hydrophobic interaction involves van der Waals forces it can be expected to contribute to retention behaviour in a lot of systems, even on media not purposely designed for HIC. It has been compared with salting-out of hydrophobic proteins from an aqueous solution and experiences from such systems may be utilized in the design of a separation by HIC.

5.2.1 Separation principle

The separation principle of HIC is not yet fully elucidated (24). In one theory it is assumed that the ordered layer of water molecules surrounding the hydrophobic ligand and the interacting hydrophobic patch of the protein is released at adsorption. This will lead to an increase in the entropy of the system which will result in a net decrease in free energy of the system (9, 12). Another explanation is based on the solvophobic theory where the influence of salt on the surface tension of the solvent is the dominating effect (7). A third theory, which is supported by thermodynamic measurements, points at the surface interactions between protein and the ligand as providing the driving force (2, 10). All theories support that interaction is related to the hydrophobic surface area of the protein and that it is increased by high ionic strength and high temperature. On the other hand, some observations are yet unexplained by the current theories and therefore purification of proteins using HIC is still to a large degree founded on empirical ground.

5.2.2 Critical factors

Since hydrophobic interaction is carried out at high ionic strength the risk for protein precipitation in the system or on the column must be considered. Likewise, an adsorbent yielding too strong a hydrophobic interaction may result in conformational changes of the protein and losses of activity. This also makes the time factor important to control. The type of ligand has a decisive influence on the retention. It is, generally, recommended to use the most hydrophobic ligand that is compatible with the protein. This will allow a low ionic strength to be used during the adsorption as compared to when using a weaker hydrophobic ligand. As for other adsorptive modes the choice of start buffer (i.e. ionic strength) to promote quantitative sorption, instead of unpredictable retardation is important and the slope of the gradient, the protein load, particle size, column length, and flow rate are listed in decreasing importance for the desorption process. Though it may be expected that the type of ions should have a strong influence it is possible to use a few combinations of ions and adsorbents to solve the majority of protein purification situations (25). Hydrophobic interaction generally decreases with decreasing temperature. Therefore, applying a refrigerated sample directly to a column may result in loss in binding or early elution (26).

5.2.3 Optimization of binding and elution conditions

The binding and elution conditions in HIC can be evaluated in the same way as described for IEC (see *Protocol 1*). Thus, for the adsorption step the ionic strength of the binding salt (e.g. ammonium sulfate or potassium sulfate) is varied using a buffer substance at a pH that will promote solubilization of the protein (ammonium cannot be used at basic pH due to formation of ammonia gas). The pH is generally not varied. The interaction strength of adsorbents generally varies in the order phenyl > octyl > butyl. A recommendation is to use a strong adsorbent (e.g. phenyl) to allow for lower concentration of adsorption buffer (to reduce the risk for protein precipitation). From the binding studies information on the desorption step is also obtained. If the protein is strongly bound to the adsorbent, it may be necessary to add ethylene glycol or methanol to the desorption buffer or to try a weaker adsorbent. Additives such as urea, guanidine hydrochloride, and detergents that may be expected to influence protein structure can be used but are generally not recommended. The procedure (i.e. *Protocol 1*) can also be used for the determination of protein solubility in various salts at different ionic strengths. This information is generally sufficient for direct optimization by chromatography. If different media are to be screened the following conditions may be used as a starting point:

(a) **Adsorption buffer**. 0, 0.2, 0.5, 1.0, 1.5, and 2.0 M potassium sulfate or ammonium sulfate (the latter is a stronger adsorption buffer in HIC) in 0.02 M phosphate buffer pH 7.

(b) **Desorption buffer**. 0.02 M phosphate buffer pH 7. It is important to check the recovery of active material after the desorption step. Investigating the precipitation or aggregation as a function of time (e.g. by measuring the turbidity) will yield important information for the further optimization work.

Other salts that may be tested for binding can be found in the Hofmeister series for the ability of ions to precipitate proteins (24). Increasing precipitation is found in the order: $SCN^- < I^- < ClO_4^- < NO_3^- < Br^- < Cl^- < CH_3COO^- < SO_4^{2-} < PO_4^{3-}$ and $Ba^{2+} < Ca^{2+} < Mg^{2+} < Li^+ < Cs^+ < Na^+ < K^+ < Rb^+ < NH_4^+$.

Since hydrophobic interaction is influenced by temperature, all solutions should be allowed to adopt ambient temperature prior to the experiments.

In case a more hydrophobic protein is present in the sample, it may be a good strategy to try to remove this component by precipitation or adsorption on an HIC medium under conditions where the target protein does not bind. These conditions may be examined by following *Protocol 1*.

5.2.4 Experimental set-up and optimization

When the binding conditions have been established, the experimental set-up is assembled as outlined in *Protocol 2*. Typical starting conditions for the optimization are given in *Table 3*. It is important to ensure that the salt content of the equilibration and wash buffers are sufficient to retain the protein and not just retard it, yielding irreproducible results. Since the protein may precipitate at too high a salt content it has been suggested to mix the protein solution and concentrated binding buffer on line prior to sample application (27). The effect of hold-up times in the system (especially for large scale systems) on protein precipitation or aggregation must also be considered when using extreme conditions. However, in general, conditions that eliminate the risk of precipitation should be used to create a robust separation method.

As stated above, the optimization should generally start by evaluating the most hydrophobic gel first. The reason is that this will reduce the risk of protein losses in the system since a lower ionic strength of the adsorption buffer can be used (and also from economic and environmental considerations). This strategy is possible since the effect of different types of ligands may be regulated by a choice of salt (i.e. strong ligand–weak salt and vice versa) and that the separation factor using different salts may be regulated by the gradient (25).

5.3 Reversed-phase chromatography

RPC is carried out in an aqueous solution containing a water miscible organic solvent, or modifier, of varying concentration. The adsorbent is surface modified to carry organic ligands (e.g. alkyl chains of different lengths C1–C18) or composed of an organic polymer providing the apolar surface required for the separation. Carbon chain-lengths used for the separation of proteins are typically C4–C8 (i.e. as given by butyl to octyl substituents). The interaction with apolar solutes is strong and requires the use of high concentrations of organic modifier, such as acetonitrile, to promote desorption of the solute.

5.3.1 Separation principle

The hydrophobic patches of the protein interact with the apolar surface of the RPC medium causing retention of the protein. The exact mechanism of this interaction is not understood in detail and different theories have been proposed and supported by experimental results (28). These include solvophobic effects where the solute is excluded from the polar mobile phase (i.e. thus a solvent effect), a partitioning effect where the solute is preferentially partitioning into the organic layer of the RPC medium, and finally an adsorptive effect where the solute is simply adsorbed to the surface of the RPC medium (28–30). It is reasonable to assume that all these effects contribute in various degrees to the retention of different types of molecules. Thus, it has been found that small molecules are probably subject to partitioning while larger ones, such as proteins are probably only retained by adsorption. From this it may be expected that van der Waals forces are important, and that hydrogen bonding in the solvent may contribute (i.e. to the solvophobic effect). The high surface density of RPC media and the preferential accumulation of the organic modifier on the RPC surface provides a strong apolar environment that may eventually cause re-arrangement and denaturation of proteins. This is a reason why RPC has not been employed as a general tool for preparative purification of proteins. RPC differs from HIC in the density of surface coverage, which is an order of magnitude higher than for HIC, and also in the general use of organic solvents in the mobile phase in order to desorb the protein. However, as indicated above, the basic molecular interactions are very similar to HIC, and RPC may conceptually be regarded as a strong type of HIC or vice versa.

5.3.2 Critical factors

The pore size of a RPC media for proteins should be at least 300 Å, in contrast to 100 Å commonly used for peptides. A mobile phase that dissolves the protein and additives suppressing ionic sites on the protein may be needed. It may be noted that very lipophilic proteins may not be suitable for separation by RPC since they may adsorb too strongly. For this type of molecules a weak HIC adsorbent should be tested. The column length is generally not critical for the separation due to the small elution window of proteins. As for IEC the multiple point interaction of proteins will normally require a gradient elution and de-sorption over a very narrow change in mobile phase composition (sometimes referred to as the on–off mechanism). As with other adsorptive modes, the choice of start buffer to promote quantitative sorption, instead of unpredictable retardation, is important and the slope of the gradient, the protein load, particle size, column length, and flow rate are important in the purification process. Highly charged proteins do not normally chromatograph well in RPC and it may be needed to decrease the charge by ion-pairing (see Section 3). It is also im-portant to select additives and organic modifiers that are UV transparent (e.g. acetonitrile).

5.3.3 Optimization of binding and elution conditions

The binding conditions are usually not critical, as long as the concentration of organic modifier is below around 1–2%. When using silica based RPC media the pH is normally neutral or acidic to avoid dissolution of the base matrix. Polymer based media do not suffer from this limitation. Experimental conditions for adsorption and desorption may be screened according to *Protocol 1*. It is im-portant to control the recovery of active material after the desorption step. If different media are to be screened the following conditions may be used as a starting point:

(a) **Adsorption buffer.** 0.1 M phosphate buffer pH 7.

(b) **Desorption buffer.** 0%, 5%, 10%, 20%, 40%, and 60% acetonitrile in 0.02 M phosphate buffer pH 7. It is important to check the recovery of active material after the desorption step! Investigating the precipitation or aggre-gation as a function of time will yield important information for the further optimization work!

5.3.4 Experimental set-up and optimization

RPC differs from IEC and HIC in that it is mainly used for polishing (see Section 5.6). Typical experimental conditions for this application in large scale are:

(a) Particle size 15–30 μm, polymeric media or octadecyl (C8) bonded silica.

(b) Bed length 10–20 cm.

(c) Fluid velocity 100 cm/h.

(d) Sample residence time > 5 min.

(e) Gradient of 0–100% acetonitrile in 20 column volumes.

(f) Sample load of 30% of maximum load (20).

The buffer is optional, as long as the salt is compatible with the organic modifier. Organic modifiers that may be tested are listed in *Table 1*. The elution strength increases in the order methanol < acetonitrile < propanol. However, the high viscosity of propanol limits its usefulness. The elution of proteins in RPC is preferentially performed by gradient which allows the separation of very minute differences in polypeptide composition.

The experimental set-up of an RPC system follows the outline of a conventional gradient system. However, all wettable parts must be resistant to organic solvents used. Optimization of the separation is carried out according to *Protocol 2*. The gradient may be followed by conductivity and the proteins detected by UV, provided a UV-transparent organic modifier is used. The cost of organic solvents, the need for explosion-proof facilities, and the disposal of organic solvent are factors that hamper the use of large scale conventional RPC.

5.4 Charge-transfer chromatography

Many chromatographic interactions involve transfer of electrons or interactions between electron orbitals. However, in some cases this interaction is specific (i.e. it requires certain types of charged surfaces and/or sites on the target molecule). Examples of these types of specific electron donor–acceptor interactions are given by IMAC and hydroxyapatite chromatography. Other types of electron donor acceptor interactions that have been noted include aromatic interaction and hydrogen bonding. However these two interactions have not hitherto been used for protein purification except in combination with other mechanisms (e.g. in dye-affinity chromatography). Aromatic adsorption have been used for separation of organic solutes (31). Hydrogen bonding has been employed for preparative purification of a large variety of plant extracts and natural polar organic molecules using non-aqueous solvents (32). Immobilized metal-affinity chromatography is currently of great interest (e.g. for purification of poly-histidine tagged recombinant proteins) and is treated in some detail below while the use of hydroxyapatite chromatography is briefly described.

Amino acids such as histidine, tryptophan, tyrosine, and phenylalanine can act as electron donors when displayed on the surface of proteins. These may interact with electron acceptors, such as transition metal ions. This is the separation mechanism underlying immobilized IMAC first described by Porath and co-workers (33). The metal ion, e.g. Cu^{2+} or Ni^{2+}, is immobilized by using a ligand having chelating properties, such as iminodiacetic acid (IDA) or nitrilotriacetic acid (NTA).

5.4.1 Separation principle

Separation in IMAC is caused by the interaction between electron donor groups on the surface of the protein and electron acceptor groups, attached to the stationary phase, via a coordination complex (34, 35). The electron acceptor group is commonly a metal ion such as Cu^{2+}, Ni^{2+}, and Zn^{2+}. Copper ions are used together with IDA and nickel ions together with IDA or NTA. The ion-ligand chelating complex is strong and up to 20 runs may be performed before the adsorbent (e.g. IDA) needs to be re-loaded with Ni^{2+} (36). The protein–metal ion coordination complex is allowed to be formed at an intermediate pH that will assure deprotonation of the interacting amino acids of the protein. The interaction is particularly strong with histidine residues and recombinant proteins are often expressed fused to a poly-His tag (commonly 6His) to allow for a selective IMAC step for purification. The protein capacity for capture of histidine tagged proteins from cell lysate is typically 6 mg/ml for Ni–IDA and 5 mg/ml for Ni–NTA (36). The desorption is performed by competition using a strong electron donor such as imidazole or decreasing the pH to protonate the interacting amino acids (e.g. at pH 4).

5.4.2 Critical factors

In IMAC it is critical that the buffers are free from complexing agents (e.g. EDTA) that otherwise will strip the ion from the adsorbent. Different metal ions have different affinities for proteins and the binding strength generally increases in the order $Ca^{2+} < Co^{2+} < Ni^{2+} < Zn^{2+} < Cu^{2+}$ (37). The pH should be neutral or basic during the adsorption step. The ion strength of the buffers are 0.5–1 M to reduce ion exchange effects.

5.4.3 Optimization of binding and elution conditions

Optimization involves the selection of type of chelating gel (i.e. the chelating functionality) and also the selection of metal ion (though Ni^{2+}, Zn^{2+}, and Cu^{2+} seem to be the most popular for laboratory work). The upper part of the gel must be charged with ions prior to use. The binding and elution conditions are evaluated according to the procedure outlined in Protocol 2:

(a) **Regeneration solution**. Charge the gel by passing a solution of 1–5 mg/ml of $NiSO_4$, $CuSO_4$, or $ZnCl_2$ through the column. The entire column should not be charged which may be difficult to avoid. Alternatively, the column is charged and then a specific volume of a solution containing 0.05 M EDTA is pumped into the column from the bottom end piece. The volume should be equal to 40% of the bed volume for Cu^{2+} and 15% for Zn^{2+}. This will leave the corresponding volume in the column free for capture of leaking metal ions during the run.

(b) **Equilibration solution**. Run two column volumes of binding buffer through the column.

(c) **Binding buffer**. 20 mM phosphate buffer pH 7 containing 0.5–1 M NaCl.

(d) **Wash buffer** is the same as binding buffer (use 5–10 CV for washing unbound substances).

(e) **Elution buffer**. 20 mM phosphate buffer pH 7 containing 0.5–1 M NaCl and 50 mM imidazole or 1 M NH_4Cl.

(f) **Cleaning solution**. Strip the column of metal ions by running two column volumes of 0.05 M EDTA followed by a wash of two column volumes 1 M NaCl. With Ni^{2+} and Cu^{2+} it is possible to run several samples prior to regeneration of the column.

The experimental set-up may be very simple, as generally is the case for systems of high association constants. For instance, when purifying a poly-His tagged protein the association constant may be as high as 10^{13} (at pH 8) and other solutes are readily washed through. The high association constant allows the use of detergents, high salt concentration (e.g. 2 M KCl), 6 M guanidine, or 8 M urea to desorb other proteins, provide reducing conditions (34), and allow purification of denatured proteins from solubilized inclusion bodies.

5.4.5 Hydroxyapatite

Hydroxyapatite is composed of crystals of $Ca_5(PO_4)OH$ and acts as an anion exchanger due to the surface coverage of Ca^{2+} (38). The selectivity of hydroxyapatite seems to be due to interaction with acidic groups, especially carboxylate, of the proteins. This interaction is not affected by high content of sodium chloride or ammonium sulfate which renders hydroxyapatite chromatography suitable for purification of proteins in the presence of these ions, such as after an ion exchange step or precipitation of hydrophobic contaminants (38). Elution is carried out by a step or gradient of phosphate buffer (e.g. up to 350 mM). The separation mechanism in hydroxyapatite chromatography is not fully understood, but it is evident that it displays different selectivity as compared to other modes of chromatography and is therefore employed to separate proteins on an empirical basis. Hydroxyapatite has certain drawbacks, one being the fragility of the material (38). Other types of materials such as fluoroapatite are claimed to give similar separation properties as hydroxyapatite.

5.5 Covalent chromatography

5.5.1 Separation principle

Covalent chromatography is predominantly used for separating thiol-containing proteins from non-thiol containing proteins early in the purification scheme. The principle is based upon the formation of a S–S bond between the protein and the ligand, washing away the impurities and the cleavage of the bond to elute the protein (33). The eluting agent (e.g. cysteine) may then be removed by desalting (34).

5.5.2 Critical factors

For the reaction to take place the wanted thiol groups of the protein need to be in a reduced form and access of thiol containing impurities (e.g. glutathione) must be removed, for example by desalting on Sephadex® G-25, prior to the covalent chromatography step. This is important due to the low capacity (e.g. 5 mg/ml) of adsorbents for covalent chromatography (39). Due to the slow reaction kinetics the contact time is a critical factor. The chromatography procedure is carried out by separate steps and to prevent oxidation of the thiols the sample buffer must be de-aerated.

5.5.3 Optimization of binding and elution conditions

Preparation of the sample involves the activation of thiol groups and desalting to remove reducing and activation agents. The sample is applied to a thiol containing gel, typically in a 0.1 M phosphate or Tris buffer pH 7–8, containing 0.5–1 M NaCl and 1 mM EDTA (to complex heavy metal ions that otherwise will interfere). The protein is allowed to bind to the gel (this may require a contact time of at least 20 min). After washing the impurities off the gel, the protein is eluted, either by reducing conditions (e.g. by adding 2-mercaptoethanol) or by competitive elution using 25 mM L-cysteine. The amount of reactive thiols can be determined by the reaction with 2,2'-dipyridyl disulfide. This will give information about the amount of gel that will be required for quantitative adsorption of the sample.

5.5.4 Experimental set-up and optimization

An experimental set-up for covalent chromatography is shown in *Protocol 3*.

Protocol 3

Typical procedure for covalent chromatography

The following procedure illustrates the basic steps in purification of thiol containing proteins by covalent chromatography (34).

Equipment and reagents

- Chromatography system
- Peristaltic pump
- Fraction collector
- Recorder
- Empty column, typically 20 cm long (e.g. Pharmacia column XK 16/20, maximum bed volume 40 ml). The diameter is selected according to the capacity needed for the process to give the total amount of gel required. The column should be equipped with a packing reservoir capable of handling a volume equal to the packed bed volume.

- Detector (e.g. UV detector at 343 nm to follow the coupling reaction and 280 nm to follow the protein concentration)
- Injection device (may be a loop injector for application of small volumes or a pump for application of large volumes)
- Thiol adsorbent (e.g. Thiopropyl Sepharose® 6B, freeze-dried)
- Equilibration buffer: 0.1 M Tris–HCl or phosphate, pH 7.5, 0.5 M NaCl; if contamination by heavy metals is suspected, add EDTA to a final concentration of 1 mM (de-aerate the buffer)

Protocol 3 continued

- Elution buffer: add L-cysteine to the equilibration buffer to a final concentration of 25 mM

- Cleaning solution: 0.1% Triton X-100 in equilibration buffer

- Desalting column, Pharmacia PD-10, Sephadex G-25 M

- Regeneration solution: saturated solution of 2,2'-dipyridyl disulfide. This is prepared by adding 40 mg disulfide to 50 ml of 25 mM borate buffer pH 8.0 and stirring the solution for several hours. Insoluble material is removed by filtration and the pH adjusted.

A. Preparation of the column

1 Weigh out the required amount of thiol adsorbent. 1 g of freeze-dried Thiopropyl Sepharose 6B yields 3 ml swollen gel volume. The amount is calculated from the content of thiol groups of the sample (see part B). It has been estimated that practical capacities of gels may be as low as 0.1 μmol/ml (39).

2 Add equilibration buffer to the dry adsorbent, using 200 ml buffer/g adsorbent and let the gel swell for 15 min at room temperature. Decant excess liquid to yield a 50% gel slurry.

3 Pack the gel into the column according to the manufacturer's instruction or at a flow rate of 1.5 times the maximum flow rate used during chromatography (see below).

B. Preparation of the sample

1 Desalt the sample solution by gel filtration on a PD-10 column using equilibration buffer.

2 Determine the content of reactive thiol groups. Dilute the desalted sample solution to 1 mg/ml with equilibration buffer. Add 0.2 ml regeneration solution to 2 ml of the diluted sample solution and let the reaction take place for 5 min. Measure the absorbance of released 2-thiopyridone at 343 nm and calculate the number of moles from $\varepsilon = 8.08 \times 10^8$ M^{-1} cm^{-1} (34). This corresponds to the number of moles of thiol groups that have reacted.

3 If the sample does not contain reactive thiol groups (e.g. as for many extracellular proteins) the sample must be prepared in the following way. Reduce intramolecular disulfide bonds with 25–50 mM 2-mercaptoethanol for 5 min. Desalt the solution according to step 1. Add regeneration solution to give a final concentration of 0.2 mM of 2,2'-dipyridyl disulfide and let the solution react for 5 min. Desalt this solution according to step 1.

C. The chromatographic step

1 Add the sample to the column at a flow rate corresponding to a residence time of 2 h (e.g. 20 ml/h for a XK 16/20). Alternatively the sample may be applied and the flow shut off for a period of 2 h. The amount of protein that has been bound may be estimated from the amount of 2-thiopyridone being released from the adsorbent during the coupling reaction (see above).

Protocol 3 continued

2   Wash the column with two to four column volumes of equilibration buffer to elute all non-bound solutes. This step may be performed at a higher flow rate (e.g. two to four times the flow rate using during the sample application).

3   Desorb and collect the target protein by running the elution buffer through the column at the same flow rate as used during sample application.[a]

4   Remove low molecular weight impurities (e.g. 2-thiopyridone) from the collected protein fraction by desalting or dialysis.

5   Regenerate the adsorbent by passing one to two column volumes of regeneration solution through the column.

6   Wash the column with two to four column volumes of elution buffer to elute all unbound solutes.

7   If necessary clean the column with one column volume of cleaning solution. Run two to four column volumes of equilibration buffer through the column.

[a] Selective elution of different types of adsorbed proteins may be achieved by applying the following elution order: (1) L-cysteine (5–25 mM), (2) reduced glutathione (0.05 M), (3) 2-mercaptoethanol (0.02–0.05 M), and (4) dithiothreitol (0.02–0.05 M) (34).

5.6 Design, optimization, and scale-up of a chromatographic purification procedure

The optimization of purification as outlined above treats a single purification step. Many times one step is not sufficient to produce a protein of required purity. Therefore a number of techniques need to be employed, and to secure a high total yield the number of steps should be kept low (e.g. three steps). By a purposeful selection of techniques the number of handling steps (e.g. conditioning steps) may be minimal. A general recommendation is to combine IEC–HIC–GF, HIC–IEC–RPC, or HIC–IEC–GF where GF stands for gel filtration (size exclusion chromatography).

The different steps are generally designed to fulfil different purposes. The first step serves to rapidly isolate or capture the target protein from the feed to stabilize the product, concentrate the product, and clear the feed from detrimental impurities (e.g. proteases). Crude feed (i.e. containing cells or cell debris) may be applied to a capture step without prior clarification by the use of adsorbents designed for expanded bed chromatography (40). The second step is an intermediate purification that aims at removal of the major part of non-product proteins, nucleic acids, endotoxins, and viruses. The third step is a polishing step to remove trace amounts of impurities, including structural variants of the target protein. Depending upon the feed and the requirements on the purity of the product the number of steps can be reduced or may need to be expanded. One example was given by the purification of recombinant human lysosomal alfamannosidase expressed in *Pichia pastoris* (41). The feed was clarified by ultrafiltration before the protein was captured by anion exchange chromatography

(Q Sepharose Fast Flow), purified by HIC (Phenyl Sepharose HP), and polished by gel filtration (Superdex® 200 prep grade) resulting in a final purification factor of 719. This example also demonstrates that media of large particle size are used early in the purification train when large volumes of crude feed are handled. An alternative to ultrafiltration and AIEC would have been to apply the feed directly to STREAMLINE® DEAE adsorbent (40).

Once the individual chromatographic steps have been optimized in concert to give the final purity and yield the purification scheme may be scaled. An adsorption chromatography process is scaled linearly and the following guidelines can be applied (20).

(a) **Maintain**:
- bed height (cm)
- fluid velocity (cm/h)
- sample concentration (mg/ml)
- gradient slope/bed volume (%B/ml gel)
- sample residence time (i.e. by keeping bed height and fluid velocity constant)

(b) **Increase**:
- column diameter (cm), to increase the bed volume in proportion to increase in sample volume
- sample volume in proportion to column cross-sectional area (ml/cm^2)
- volumetric flow rate in proportion to column cross-sectional area (ml/h and cm^2)
- gradient volume in proportion to column cross-sectional area (ml/cm^2)

(c) **Check**:
- reduction in supportive column wall effects (e.g. compaction of bed)
- sample distribution system (e.g. extra column zone broadening, sample pre-mixing in HIC)
- piping (tubing) and system dead volumes
- system hold-up times (e.g. gradient formation, sample aggregation, or denaturation in HIC)

Acknowledgements

The helpful comments to this manuscript from colleagues at APBiotech is gratefully acknowledged. In this respect John Brewer deserves special thanks.

The following trademarks mentioned in the text are owned by APBiotech; Sephadex, Sepharose, Superdex, Mono S, PolyBuffer, ÄKTA, HiTrap, and STREAM-LINE.

References

1. Karger, B. L., Snyder, L. R., and Horvath, C. (1973). *An introduction to separation science*. John Wiley & Sons Inc., New York.
2. Israelachvili, J. and Wennerström, H. (1996). *Nature*, **379**, 219.

3. Mant, C. T., Zhou, N. E., and Hodges, R. S. (1992). In *Chromatography* (5th edn) (ed. E. Heftmann), p. B137. Elsevier, Amsterdam.

4. Creighton, T. E. (1993). *Proteins, structures and molecular properties*. W. H. Freeman and Company, New York.

5. Gómez, J., Hilser, V. J., Xie, D., and Freire, E. (1995). *Proteins*, **22**, 404.

6. Pharmacia Biotech AB, Uppsala. (1995). *Ion exchange chromatography, principles and methods*.

7. Melander, W. R., Rassi, Z. E., and Horvath, C. (1989). *J. Chromatogr.*, **469**, 3.

8. Ståhlberg, J., Jönsson, B., and Horváth, C. (1992). *Anal. Chem.*, **64**, 3118.

9. Perkins, T. W., Mak, D. S., Root, T. W., and Lightfoot, E. N. (1997). *J. Chromatogr.*, **766**, 1.

10. Privalov, P. L. and Gill, S. J. (1989). *Pure Appl. Chem.*, **61**, 1097.

11. Bristow, P. A. (1976). *LC in practice*. HETP, Wilmslow, England.

12. Tanford, C. (1980). *The hydrophobic effect, formation of micelles and biological membranes*. John Wiley & Sons, New York.

13. Janson, J.-C. and Rydén, L. (ed.) (1989). *Protein purification, principles, high resolution methods and applications*. VCH Publishers Inc., New York.

14. Lagerlund, I., Larsson, E., Gustavsson, J., Färenmark, J., and Heijbel, A. (1998). *J. Chromatogr.*, **796**, 129.

15. Williams, A. (1995). In *Current protocols in protein science*, Vol. 1 (ed. J. E. Coligan, B. M. Dunn, H. L. Ploegh, and P. T. Wingfield), p. 8.2.17. John Wiley & Sons, Inc., New York.

16. Gallant, S. R., Kundu, A., and Cramer, S. M. (1995). *J. Chromatogr.*, **702**, 125.

17. Dolan, J. W. and Snyder, L. R. (1987). *LC/GC*, **5**, 970.

18. Gill, D. S., Roush, D. J., Shick, K. A., and Willson, R. C. (1995). *J. Chromatogr. A*, **715**, 81.

19. Malmquist, G. and Lundell, N. (1992). *J. Chromatogr.*, **627**, 107.

20. Sofer, G. and Hagel, L. (1997). *Handbook of process chromatography, a guide to optimization, scale-up and validation*. Academic Press, London.

21. Stafström, N. and Malmquist, G. (1996). Fast and convenient pH optimization in ion exchange chromatography using an on-line buffer preparation unit. Presentation at *ISPPP '96, Luxemburg*.

22. Hutchens, T. W. (1989). In *Protein purification, principles, high resolution methods and applications* (ed. J.-C. Janson and L. Rydén), pp. 149–74. VCH Publishers Inc., New York.

23. Logan, K. A., Lagerlund, I., and Chamow, S. M. (1999). *Biotechnol. Bioeng.*, **62**, 208.

24. Pharmacia Biotech AB, Uppsala. (1993). *Hydrophobic interaction chromatography, principles and methods*.

25. Gagnon, P. and Grund, E. (1996). *BioPharm*, **9**, 54.

26. Gagnon, P., Grund, E., and Lindbäck, T. (1995). *BioPharm*, **8**, 36.

27. Gagnon, P., Grund, E., and Lindbäck, T. (1995). *BioPharm*, **8**, 21.

28. Pharmacia Biotech AB, Uppsala. (1996). *Reversed phase chromatography, principles and methods*.

29. Dorsey, J. G. and Cooper, W. T. (1994). *Anal. Chem.*, **66**, 857A.

30. Carr, P. W., Tan, L. C., and Park, J. H. (1996). *J. Chromatogr.*, **724**, 1.

31. Pharmacia Biotech. (1990). *Sephadex LH-20, chromatography in organic solvents*.

32. Henke, H. (1995). *Preparative gel chromatography on Sephadex LH-20*. Hütig, Heidelberg, Germany.

33. Porath, J., Carlsson, J., Olsson, I., and Belfrage, G. (1975). *Nature*, **258**, 598.

34. Pharmacia Biotech AB, Uppsala. (1993). *Affinity chromatography, principles and methods*.

35. Kågedal, L. (1989). In *Protein purification, principles, high resolution methods and applications* (ed. J.-C. Janson and L. Rydén), pp. 227–51. VCH Publishers Inc., New York.

36. Heijbel, A. (1995). Personal communication.

37. Petty, K. J. (1995). In *Current protocols in protein science*, Vol. 1 (ed. J. E. Coligan, B. M.

Dunn, H. L. Ploegh, and P. T. Wingfield), pp. 9.4.1.–9.4.16. John Wiley & Sons, Inc., New York.

38. Karlsson, E., Rydén, L., and Brewer J. (1989). In *Protein purification, principles, high resolution methods and applications* (ed. J.-C. Janson and L. Rydén), pp. 138–9. VCH Publishers Inc., New York.

39. Rydén, L. and Carlsson, J. (1989). In *Protein purification, principles, high resolution methods and applications* (ed. J.-C. Janson and L. Rydén), pp. 252–74. VCH Publishers Inc., New York.

40. Pharmacia Biotech AB, Uppsala. (1997). *Expanded bed adsorption, principles and methods.*

41. Liao, Y.-F., *et al.* (1996). *J. Biol. Chem.*, **271**, 28348.

Chapter 8

Chromatography on the basis of size

P. Cutler

SmithKline Beecham Pharmaceuticals, New Frontiers Science Park (North), Third Avenue, Harlow, Essex CM19 5AW, UK.

1 Introduction

Chromatography has been employed for the separation of proteins and other biological macromolecules on the basis of molecular size since the mid 1950s when Lathe and Rutheven employed modified starch as a media for separation (1). Porath and Flodin developed the technique further using cross-linked dextran and coined the term *gel filtration* (2). Some confusion over nomenclature has been created by the term *gel permeation*, used to describe separation by the same principle in organic mobile phases using synthetic matrices (3). It is now generally agreed that the terms gel filtration and gel permeation do not accurately reflect the nature of the separation. *Size exclusion chromatography* (SEC) has been widely accepted as a universal description of the technique and in line with the IUPAC nomenclature this term will be adopted (4). The historical development of SEC for protein separation has been reviewed (5).

SEC is a commonly used technique due to the diversity of the molecular sizes of proteins in biological tissues and extracts. In addition to isolating proteins from crude mixtures, SEC has been employed for many roles including buffer exchange (desalting), removal of non-protein contaminants (DNA, viruses) (6), protein aggregate removal (7), the study of biological interactions (8), and protein folding (9–11).

1.1 Principle

1.1.1 The mechanism of separation

The principle of size exclusion is based on a solid phase matrix consisting of beads of defined porosity which are packed into a column through which a mobile liquid phase flows (*Figure 1*). The mobile phase has access to both the volume inside the pores and the volume external to the beads. Unlike many other chromatographic procedures size exclusion is not an adsorption technique. Separation can be visualized as reversible partitioning into the two liquid

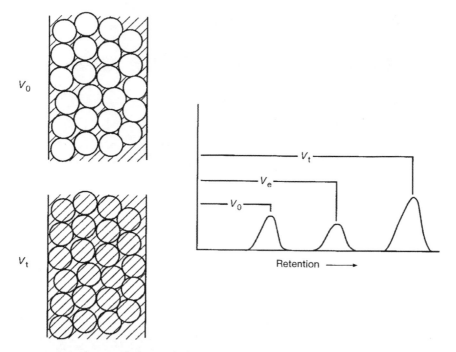

Figure 1 Principle of SEC. In SEC solutes are separated according to their molecular size or hydrodynamic volume. Large molecules are unable to enter the pores of the gel particles and are eluted in the void volume (V_0). Small molecules are free to enter the pores and are eluted in the total volume (V_t). Molecules with a molecular size within the separation range of the matrices are fractionally excluded with a characteristic elution volume (V_e).

volumes. The elution time is dependent upon an individual protein's ability to access the pores of the matrix. Large molecules remain in the volume external to the beads as they are unable to enter the pores. The resulting shorter flow path means that they pass through the column relatively rapidly, emerging early. Proteins that are excluded from the pores completely, elute in the void volume, V_0. This is often determined experimentally by the use of a high molecular weight component such as blue dextran or calf thymus DNA. Small molecules which can access the liquid within the pores of the beads are retained longer and therefore pass more slowly through the column. The elution volume for material which is free to move in and out of the pores is the total volume, V_t. This represents the total liquid volume of the column and is often determined by small molecules such as vitamin B12. The actual protein or peptide used to calculate V_t and V_0 will depend on the separation range of the particular matrix employed. The difference in volume between V_0 and V_t is the volume within the pores known as the stationary phase volume (or 'partially stagnant' phase) and is designated V_i, Equation 1:

$$V_i = V_t - V_0 \qquad [1]$$

Intermediate sized proteins will be fractionally excluded with a characteristic elution volume, V_e, which lies between V_0 and V_t. V_e is normally calculated as the volume taken from the point of sample injection to the apex of the eluted peak. A distribution coefficient K_{av} can be determined for each protein, Equation 2:

$$K_{av} = \frac{(V_e - V_0)}{(V_t - V_0)} \tag{2}$$

K_{av} is not a true partition coefficient (K_d) as $V_t - V_0$ includes the volume occupied by the solid matrix component which is inaccessible to all solutes. However, for any given matrix the relationship between K_{av} (which is readily determined) and K_d is constant. The value V_i can be substituted for $(V_t - V_0)$ in Equation 2, however in practice, V_t and V_0 are readily calculated as described above and V_i can only readily be derived from Equation 1. Various methods exist for defining the mechanisms by which separation occurs but they are all based on quantitating the proportion of V_i available to a given solute (12, 13).

The K_{av} of a protein does not directly relate to the molecular weight (M_r), but to the hydrodynamic volume or radius of gyration (R). Whilst there is a relationship between K_{av} and the log of the molecular weight (14), the value of R is also dependent upon the tertiary structure of the protein as indicated in *Figure 2*. The relationships between M_r and R vary such that $R \propto M_r^{1/3}$ for compact spheres, $R \propto M_r^{1/2}$ for random coil, and $R \propto M_r$ for a rod-like protein. Other factors affecting the hydrodynamic volume include the degree of hydration experienced by the protein and interactions with other solutes. Whilst this limits accurate mass determination it enables SEC to be used analytically for studying protein folding and protein interactions (9).

The main limitations of size exclusion are related to the fact that all proteins elute in one column volume or less. Therefore, in comparison with other adsorption techniques SEC offers only moderate resolution with a low capacity. In addition to the limited ability for mass determinations SEC is vulnerable to solute interactions with the matrices, leading to so-called non-ideal size exclusion effects. SEC is, however, one of the most widely used chromatographic techniques due to

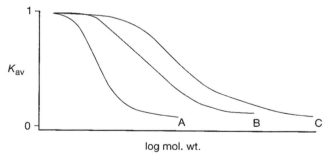

Figure 2 Effect of protein folding on retention time. Proteins display a distribution coefficient (K_{av}) which is broadly related to molecular size. Molecular size is based on molecular weight and the tertiary structure. Compact spherical proteins (C) elute later (i.e. possess a higher K_{av}) than molecules of the same molecular weight that adopt a random coil structure (B). Similarly, rod-like proteins (A) elute with yet a lower K_{av} value. Proteins with a more open structure separate with a narrower molecular weight range.

its versatility, simplicity, and cost and has the advantage of requiring little prior knowledge of the target protein to achieve a degree of purification. Unlike other techniques there is no need for harsh elution conditions and proteins are often separated under near physiological conditions. There is no reliance on the specific charge profile, hydrophobicity, or biological affinities of proteins for separation and therefore the technique has certain advantages in resolving closely related species such as aggregates or fragments. As elution is performed using isocratic mobile phases, requiring no column re-equilibration between chromatographic runs, SEC is also relatively easy to automate into multi-cycle processes.

1.1.2 Separation in terms of theoretical plates

The partitioning process occurs as the bulk flow of liquid moves down the column. As the sample is loaded it forms a sample band on the column. In considering the efficiency of the column, partitioning can be perceived as occurring in discrete zones along the axis of the column's length. Each zone is referred to as a theoretical plate and the length of the zone termed the theoretical plate height (H), This can be calculated as in Equation 3 where (L) is the length of the column in metres and N is the number of theoretical plates:

$$H = \frac{L}{N} \qquad [3]$$

The value of L is calculated directly as the bed height of the column and N is calculated from the performance of the column in terms of the elution of a particular solute peak as described in Equation 4. $W_{1/2}$ represents the width of the peak at half the maximum height, expressed in the same units as V_e (*Figure 3*):

$$N = 5.54 \cdot (V_e / W_{1/2})^2 \qquad [4]$$

The highest resolving matrices have high values for N and correspondingly low values for H. In practice, the number of theoretical plates (N) can be measured using a suitable UV absorbing material such as 1% (v/v) acetone monitored at 280 nm. Other aromatic compounds can be used such as tyrosine or benzyl alcohol.

As proteins move down the column, some band spreading is inevitable due to the nature of the mass transfer of proteins between the mobile phase and pores

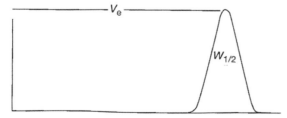

Figure 3 Calculation of theoretical plate height. The theoretical plate height (H) of a column as a measure of matrix performance. The number of theoretical plates (N) is related to the peak width at half-height ($W_{1/2}$) and the elution volume (V_e).

of the support. The causes of band spreading can however be minimized. An alternative definition of H is described in the van Deemter Equation 5 in which the value of H is related to various diffusional factors which can lead to band spreading:

$$H = A + \frac{B}{u} + C \cdot u \qquad [5]$$

A is derived from factors leading to multiple path diffusion of the solute (e.g. the average pore size diameter and flow rate). B is related to the axial diffusion factors (e.g. obstruction factors and flow rate) and is often considered negligible for macromolecules such as proteins. C is derived from non-equilibrium conditions and u is the interstitial linear velocity.

1.1.3 Operational parameters influencing separation

Several parameters affect protein separation including sample volume, flow rate, column dimensions, and the mobile phase. Successful protein separation as determined by peak resolution, protein recovery, peak capacity (the number of peaks resolvable on a given column), and reproducibility is a function of the operational selectivity of the matrix employed. Matrices are supplied in defined molecular weight fractionation ranges.

The resolution (R_S) of proteins can be calculated experimentally as in Equation 6 where ΔV_e is the baseline width of the individual peaks.

$$R_S = \frac{2 \cdot (V_{e2} - V_{e1})}{\Delta V_{e1} + \Delta V_{e2}} \qquad [6]$$

As resolution is dependent upon the number of theoretical plates available, it is a function of column length (L) such that $R_S \propto \sqrt{L}$. A doubling of the column length therefore leads to an approximate 40% increase in resolution (15). Columns are typically 70–100 cm long for low pressure separations and 25–30 cm for HPLC separations.

Two important physical parameters of the matrices effecting the resolution are the particle size and porosity. As the particle size of the matrix decreases the theoretical plate height H is decreased. Hence HPLC matrices with a 5–10 μm particle size offer higher resolution than the standard 30-100 μm low pressure particles. Although particles smaller than 5 μm may offer further increases in performance, problems with packing and shear effects make particles lower than 2 μm unlikely in the near future. Porosity of the matrix should be sufficient to maximize the value of V_i whilst maintaining the rigidity of the matrix to operating parameters.

The mobile phase flow rate is a key parameter affecting resolution. Resolution generally decreases with increasing flow rate. The optimum flow rate for low pressure columns is around 5 cm/h. Moderately higher flow rates (10–15 cm/h) are typically used with little effect on resolution. The high performance HPLC matrices with smaller particle size can be used at higher flow rates to produce resolution with shorter run times.

The viscosity of the sample should be kept to a minimum as this limits mass transfer, increasing plate height, and lowering resolution. Viscosity is dependent upon several factors including the nature of the solute, solute concentration, the mobile phase, and operating parameters such as temperature. Operation of SEC columns at ambient temperature rather than 4°C has offered improved resolution, however consideration must be given to sample stability.

1.2 Applications of size exclusion chromatography

Although size exclusion is a relatively straightforward form of chromatography several factors need to be considered to optimize resolution. Some of the properties required for a successful matrix are listed below:

- macroporous to aid partitioning
- hydrophilic and non-ionic to avoid matrix–protein interaction
- spherical and uniform in particle size and pore diameter
- chemically stable to solvents and operational pH
- rigid to withstand operating pressure
- non-biodegradable
- inexpensive

Whilst the above properties are important generic characteristics, potentially the most critical parameter is the matrix selectivity. Some matrices offer a wide range of molecular weight separations and others are high resolution matrices with a narrow range of operation (*Table 1*). The matrix selected for a given fractionation must be determined based on the expected fractionation range of the gel under the operating parameters. An ideal matrix should have a linear and wide ranging selectivity profile to give maximum resolution. As proteins approach the exclusion limit of the column the mass transfer becomes less efficient and so the higher molecular weight material tend to display more band spreading. In addition the later eluting peaks display more tendency towards diffusion leading to peak broadening. Ideally the protein of interest should elute on the linear part of the selectivity profile. Recent developments in media design have focused on producing a narrow pore size distribution for high resolution and small particle sizes to yield efficiency at high flow rates. Although traditionally SEC has been employed for the separation of proteins in the range 30–300 kDa, there is renewed interest in the ability of SEC to be used for separating peptides (16). Matrices such as Superdex Peptide (Pharmacia) have a fractionation range of 100–7000 Da.

1.2.1 Preparative size exclusion chromatography

The performance and resolution of preparative size exclusion has been enhanced by the development of newer matrices with improved properties. In addition to cross-linked dextran gels (e.g. Sephadex), polystyrene-based matrices

Table 1 Commonly used preparative size exclusion matrices

Supplier	Matrix	Composition	Size range[a,b]		Stability	
			Minimum	Maximum	pH	Organics
Low/medium pressure						
Pharmacia	Sephadex	Dextran	1000–5000	5000–600 000	2–10	
	Sephacryl	Dextran/bisacrylamide	1000–100 000	500 000 to > 100 000 000	3–11	
	Sepharose CL	Agarose	10 000–4 000 000	70 000–40 000 000	3–13	
	Superose	Agarose	1000–300 000	5000–5 000 000	3–12	
	Superdex	Agarose/dextran	100–7000	10 000–600 000	3–12	
Bio-Rad	Bio-Gel P	Polyacrylamide	100–1800	5000–100 000	2–10	
	Bio-Gel A	Agarose	10 000–500 000	100 000–50 000 000	4–13	
Toso Haas	Toyopearl HW	Methacrylate	100–10 000	400 000–30 000 000	1–14	Yes
Amicon	Cellufine	Cellulose	100–3000	10 000–3 000 000	1–14	
Biosepra	Ultrogel AcA	Acrylamide/agarose	1000–15 000	100 000–1 200 000	3–10	
	Trisacryl GF	Acrylamide	300–7500	10 000–15 000 000	1–11	
Merck	Fractogel TSK	Polymeric	100–10 000	500 000–50 000 000	1–14	
HPLC						
Toyo Soda	TSK-SW	Silica	5000–150 000	20 000–10 000 000	2–7	
	TSK-PW	Polymeric	100–2000	10 000–200 000	2–12	Yes?
Bio-Rad	Bio-Sil SEC	Silica	5000–100 000	20 000–1 000 000	2–8	Yes
Dupont	BioSeries GF	Zorbax®[c]	10 000–250 000	25 000–900 000	3–8.5	
Shodex	Asahipak	Polyinyl alcohol	100–3000	10 000–10 000 000	2–9	Yes
Phenomenex	BioSep-SEC-S	Bonded silica	1000–200 000	20 000–3 000 000	2.5–7.5	Yes
	Polysep-GFC-P	Polymeric	100–2000	10 000–10 000 000	3–12	
Synchrom	Synchropak GPC	Bonded silica	300–30 000	25 000–50 000 000	?	
Shodex	OHpak	Polyhydroxymethacrylate	100–1000	10 000–20 000 000	?	
Waters	Ultrahydrogel	Polyhydroxymethacrylate	100–5000	100 000–7 000 000	2–12	
Macherey-Nagel	Nucleogel SFC	Polymeric	100–100 000	100 000–20 000 000	1–13	Yes

[a] Size range represents the maximum and minimum separation range for the class of matrices.

[b] Range is given in general as estimates for globular proteins in aqueous buffers as recommended by the manufacturer.

[c] Zorbax® is a trademark of Dupont.

have been developed for the use of size exclusion in non-aqueous conditions. Polyacrylamide gels such as the Biogel P series (Bio-Rad) are particularly suited to separation at the lower molecular weight range due to their micro-reticular structure. Matrices such as Sepharose CL-6B and Sephacryl S-500 HR (Pharmacia) can be used to separate large macromolecules, particles, viruses, DNA, etc. More recently higher performance composite matrices have been developed. These include the Superdex gels (Pharmacia) with dextran chains which determine selectivity, chemically bonded to a highly cross-linked agarose and Ultrogel AcA gels (BioSepra) a composite of polyacrylamide and agarose (*Table 1*).

Size exclusion is commonly used as a final polishing step when impurities are low in number and the target protein has been purified and concentrated by earlier chromatography steps. Size exclusion is particularly suited for the resolution of protein aggregates from monomers (7, 17). Aggregates are often formed as a result of the purification procedures used and size exclusion can also facilitate buffer exchange mechanism into the desired solution.

Preparative separations can also be used as an initial purification step allowing quick removal of impurities from crude samples (18). This is common in membrane protein purification where concentration techniques are not readily available and the material will be progressively diluted during the purification scheme (19). SEC has been used as a method of detergent exchange in membrane protein isolation (20).

In addition to lab scale preparation, low pressure SEC has been successfully used for process scale isolation of proteins including the desalting of dairy products (21) and the de-ethanolization of plasma proteins (22). Size exclusion has become an important element in the isolation of therapeutic grade proteins (23–25).

Ideally, on increasing scale, the column size can be increased in proportion to the sample volume, maintaining a constant bed height, sample concentration, and linear flow rate. However scale-up issues usually reflect the economic necessity to move to a larger particle size. Several matrices are available as analytical and preparative grade gels. Superdex and Superose (Pharmacia) matrices are available in high resolution analytical grade (10–14 μm) and preparative grade (30–35 μm) with the same selectivity. Due to the loss of performance on moving to preparative grade material, resolution is maintained by a decrease in the linear flow rate leading to increased run times. Although productivity can be enhanced to a certain degree by increasing protein concentration the low sample capacity of size exclusion process-scale columns means multi-cycle processes are employed with process time minimized by optimization of sample loading and, where possible, automation.

1.2.2 Desalting (buffer exchange)

Desalting gels are used to rapidly remove low molecular weight material such as chemical reagents from proteins and nucleic acids. Desalting is used for the rapid removal of protein labels such as the radiolabel ^{125}I and the fluorescent tag fluorescein isothiocyanate (FITC). It is an efficient method of buffer exchange, to prepare material for further chromatographic steps and for freeze-drying (26). As a result of the speed of chromatographic separations, desalting chromatography is a valid alternative to buffer exchange by dialysis or ultrafiltration.

Several gels are commercially available for desalting, including Sephadex G-25 (Pharmacia) and BioGel 6 (Bio-Rad). Pre-packed columns such as the PD-10 gravity feed columns (Pharmacia), Biospin 6 centrifugal columns (Bio-Rad), the low pressure Hitrap Desalt column (Pharmacia), and the medium pressure Fast Desalting HR10/10 column (Pharamacia) are available. As the molecules to be removed are generally very small, typically less than 1 kDa, gels are used with an exclusion limit of approximately 2–5 kDa. The protein appears in the void volume (V_0) and the reagents and buffer salts are retained, appearing at, or close to, the total volume (V_t). Because of the distinct molecular weight differences between the protein and buffer salts less theoretical plates are needed in order to achieve resolution, therefore shorter columns (e.g. 15-30 cm) are run at higher loading (e.g. 20% bed volume) and at higher flow rates (e.g. 30 cm/h).

1.2.3 High performance size exclusion chromatography

High performance columns are used analytically for studying protein purity (27). Size exclusion has the advantage of simplicity however it cannot equal the resolving power of techniques such as polyacrylamide gel electrophoresis. Size exclusion in the HPLC format has been used to study protein folding and stability (10, 11), protein aggregation (28), and protein–protein interactions (8, 29). HPLC SEC has been employed for the characterization of recombinant proteins such as tissue necrosis factor and tissue plasminogen activator (30). The provision of data supporting the structural integrity of therapeutic proteins including fragmentation and aggregation of proteins is included in the Food and Drugs Administration's Points to Consider (31). HPLC has been used in preparative mode for production of up to 100 mg of protein (32). In these cases 22 mm i.d. columns are generally used at flow rates from 2–20 ml/min.

For HPLC analysis modified silica with 5–10 μm particle size is commonly used as it gives good consistency of particle and pore diameter and is rigid to the high operating pressures (33). Such matrices include the TSK (e.g. 2000SW and 3000SW) columns from Toya Soda and the Zorbax GF (e.g. 250) columns from Dupont (*Table 2*). An alternative to silica is controlled pore glass (34). The pore size of the silica can vary from 50–4000 Å but a size range of 100–300 Å is common for separation of globular proteins using a 5–10 μm particle. Resolution of globular proteins with mass difference of ± 20% is believed possible whereas the theoretical maximum for rod-like solutes is ± 10% (7). Due to the efficiency of the matrices the columns are shorter than the low pressure gels and operate at higher flow rates (300–400 cm/h equating to 1 ml/min in a standard 4.6 mm i.d. column). Matrices such as cross-linked agarose, Superose (10 μm), and Superdex (13 μm) gels supplied by Pharmacia are used analytically although they are unable to withstand the pressures attainable with other matrices.

Table 2 Mobile phases used for separation of proteins[a]

Type	Mobile phase	pH range	Comment
Aqueous			
Non-volatile	'Good' Buffers[b]	5.5–11	The ionic strength should be kept with in the range equivalent to 0.1–0.5 M NaCl. Sulfates can lead to salting-out effects. Borates can complex with glucosidic functions on some, matrices.
	50 mM sodium phosphate, 0.15 M NaCl	7.2	Physiological.
	0.1 M Tris–HCl	7–9	Good solubility particularly for RNA and DNA.
Volatile			Volatile buffers are used for freeze-drying. They are also relatively compatible with on-line mass, spectrometric techniques, but due to the low salt can lead to non-ideal size exclusion effects.
	0.1 M ammonium acetate	7–10	Good for proteins and DNA. Should be made immediately prior to use.
	0.1 M ammonium bicarbonate (untitrated)	7.9	
	20% (v/v) acetonitrile in 0.1% (v/v) trifluoroacetic acid	2.0	Commonly used for reverse phase HPLC, therefore compatible with further processing.
Detergents			Often as additions to the aqueous buffers above. Can give erroneous M_r estimations.
	0.1% (w/v) SDS		Strongly anionic. Good solubilizing agent but denaturing. Difficult to remove.
	0.2% (w/v) Triton X-100		Non-ionic. Can be used to solubilize hydrophobic proteins, whilst retaining activity.
Denaturing	6 M guanidine–HCl, 50 mM Tris–HCl	8.6	Good UV properties at 280 nm. Proteins can often be renatured on dialysis. Used to study folding.
	6–8 M urea	< 7.0	Urea is a good solubilizing agent but may lead to carbamylation of proteins/peptides.
	1 M propionic acid		
	70% (v/v) formic acid	1.0	Good solubilizing agents for proteins and peptides.
	50 mM HCl	1.4	Good for solubilizing peptides.
Organic	FACE (formic acid, acetic acid, chloroform, ethanol, 1:1:2:1)		High level of glycerol, ethylene glycol etc. can cause severe band-spreading.

[a] It is important to consult manufacturer's details with regard to the chemical resistance of individual matrices.

[b] 'Good' buffers are a range of zwitterionic buffers specifically developed for biological work and include Mes, Mops, Hepes, and Caps.

2 Materials

2.1 Equipment

The preparative separation of proteins by size exclusion is suited to commercially available standard low pressure chromatography systems available from suppliers such as Pharmacia, Bio-Rad, Gilson, and Perseptive Biosystems. Systems require a column packed with a matrix offering a suitable fractionation range, a method for mobile phase delivery, a detector to monitor the eluting proteins, a chart recorder for viewing the detector response, and a fraction collector for recovery of eluted proteins. HPLC systems are similarly available from suppliers such as Waters, Dionex, Hewlett Packard, Gilson, and Perkin Elmer (35). Specialized equipment is needed for process scale separations. These have been reviewed by several authors (36–38).

Early low pressure systems were less sophisticated with a gravimetric feed of the mobile phase from a suspended reservoir while the most modern systems (particularly HPLC) now have computers to control operating parameters and to collect and store data. The principle of separation however remains the same and high resolution is attainable with relatively simple equipment. Several factors described below are pertinent to size exclusion and are worth some consideration to optimize performance.

2.1.1 Pumps

An important factor in size exclusion is a reproducible and accurate flow rate. The most commonly used pumps for low pressure chromatography are peristaltic pumps which are relatively effective at low flow rates, inexpensive, and sanitizable. Peristaltic pumps do however create a pulsed flow and often a bubble trap is incorporated to both prevent air entering the system and to dampen the pulsing effect. More expensive yet more accurate pumps are syringe pumps such as those seen on the FPLC® system (Pharmacia) and HPLC systems offering low pulsation. Reciprocating piston pumps are accurate at low flow rates but are prone to pulsation.

2.1.2 Column

Size exclusion, unlike some commonly used adsorption methods of protein separation, is a true chromatography method based on continuous partitioning, hence resolution is dependent upon column length. Column length is a balance between resolution and run time. Preparative columns tend towards being long and thin, typically 70–100 cm long, although desalting columns tend to be shorter. In some instances the length of the column required to obtain a satisfactory separation exceeds that which can be packed into a commercially available column (> 1 m). In these cases columns can be packed in series. The tubing connecting the columns should be as narrow and as short as possible to avoid zone spreading. Due to the limit of the bed height in terms of matrix stability process scale production columns tend to be produced as stacking systems.

Essentially several smaller bed height columns are stacked vertically in series (36).

The column must be able to withstand the pressures generated during operation and be resistant to the mobile phase. Stability is an issue when operating with organic mobile phases, high salt, or denaturants such as guanidine hydrochloride. Stability to organics is generally in the order plastic < polypropylene < glass, however particular attention should be given to seals and frits and manufacturers guidelines should be observed. The use of columns with flow adapters is recommended to allow the packing volume to be varied and provide a finished support with the required minimum of dead-space. Before use all column parts should be cleaned in an appropriate solvent and the integrity of the support nets and seals confirmed.

Due to the increased efficiency of HPLC packings, analytical HPLC columns tend, typically, to be shorter than their preparative counterparts. The advantage is excellent resolution in a short run time (10–15 min). Longer HPLC columns, are however, available for many matrices and in our laboratory we have achieved good results by placing two HPLC columns in series. HPLC columns are supplied pre-packed and are best operated with a guard column (2–10 cm) to prevent column fouling.

2.1.3 Detectors

Protein elution is often monitored by absorbance in the ultraviolet range, either at 280 nm which is suitable for proteins with aromatic amino acids or at 214 nm which detects the peptide bond. Although detection at lower wavelengths may be complicated by the absorbance properties of certain mobile phases, this can be compensated by the higher sensitivity obtainable. Diode array detectors have enabled continuous detection at multiple wavelengths enabling characterization of the eluates via analysis of spectral data (11).

Due to the analytical nature of HPLC applications, extra demands are required of HPLC detectors in terms of sensitivity and stability. For this reason fluorescent detection and electrochemical detection is not uncommon. Fluorescence detection can be achieved either by direct detection of the intrinsically fluorescent tryptophan and tyrosine residues or after chemical derivitization (e.g. by fluorescein).

SEC can be used to estimate molecular weight through appropriate calibration. However, use of low angled or multi-angled laser light scattering can provide detailed and accurate mass information (39–41). In addition, mass spectrometry and viscometers can be used to obtain mass data. The use of SEC in conjunction with on-line mass spectrometry has largely been restricted to synthetic polymers, in part due to the incompatibility of common mobile phases with the mass spectrometers. The widespread introduction of matrix assisted laser desorption time of flight (MALDI-TOF) mass spectrometry has facilitated accurate off-line mass determinations in a range of mobile phases. Some work with protein and peptides in conjunction with mass spectrometry has yielded valuable information including the dimerization state of a leucine zipper peptide (42), the proteolytic processing of the opioid prohormone prodynorphin (43), the post-translational modification of viral membrane proteins (44), and the folding of human macrophage colony stimulating factor (45). In addition to non-specific on-line monitoring systems it is quite common, particularly when purifying enzymes, to make use of specific assays for individual target molecules.

2.1.4 Sample loading

Several methods exist for sample application. It is critical to deliver the sample to the top of the column as a narrow sample zone. This can be achieved by manually loading via a syringe directly on to the column, although this requires skill and practice, or applied through a peristaltic pump, although this will inevitably lead to band spreading due to sample dilution. The sample should never be loaded through a pump with a large hold-up volume such as a syringe pump or upstream of a bubble trap.

Arguably the best method of applying the sample is via a sample loop in conjunction with a switching valve allowing the sample to be manually or electronically diverted through the loop and directly on to the top of the column. In essence a valve is inserted in the tubing immediately before the column. The valve allows either direct flow from the pump to the column or diverts flow through a sample loop of a predefined volume allowing entry of the sample to the column with a minimum of dead volume. Autosamplers are in effect sophisticated loop injectors which can inject repeatedly from sample vials. These are particularly useful for automated and repetitive work such as stability testing and many are temperature controlled.

2.1.5 Fraction collectors

A key factor in preparative protein purification is the ability to collect fractions accurately. Regardless of how efficiently the column may have separated the proteins, the accurate collection of fractions is critical. For the detector to reflect as near as possible in real time the fraction collector, the volume between the detector and the fraction collector should be minimal. Systems are available which can collect directly into a 96-well microtitre plate which can prove highly convenient for assay screening.

2.2 Mobile phase

Size exclusion has the important advantage of being compatible with physiological conditions (5, 46), however the choice of mobile phase is critical. Size exclusion matrices tend to be compatible with most aqueous buffer systems even in the presence of surfactants, reducing agents or denaturing agents, with effective pH ranges of approximately 2–12. An important exception to this is the silica based matrices which offer good mechanical rigidity but low chemical stability at alkaline pHs. Some silica matrices have been coated with dextran etc. to increase the chemical stability and increase hydrophilicity. All buffers used should ideally be filtered through a 0.2 μm filter and degassed by low vacuum or sparging with an inert gas such as helium. Common mobile phases are listed in *Table 2*.

2.2.1 Aqueous phases

Aqueous buffers are used for most protein separations with the potential exception of hydrophobic proteins such as membrane proteins where detergents are needed to allow solubility and prevent aggregation. It is important to consider the stability of the sample load in both the loading buffer and the mobile phase, necessitating considerations of appropriate pH and solvent composition which may be essential to maintain the structural and functional integrity of the target molecule. In order to prevent denaturation or digestion of the protein it is prudent to consider the addition of protease inhibitors or cofactors. Samples are often loaded at relatively high concentrations. In addition to the miscibility of the sample buffer and mobile phase the proteins must be soluble at the expected concentrations. Poor chromatography can often result from the transition of a heterogeneous protein mixture into the mobile phase on the column with loss of solubility and/or stability.

2.2.2 Denaturing agents and detergents

Separation can be enhanced by the use of denaturing agents in the mobile phase. They have been used to permit the separation of distinct subunits which would otherwise interact and co-elute. Examples include the separation of immunoglobulin G (IgG) heavy and light chain in the presence of either 1 M propionic acid or 6 M guanidine hydrochloride to disrupt hydrogen bonding. The loss of conformation due to denaturing can, with appropriate calibration, give better estimations of molecular weight due to the elimination of variation due to tertiary structure in an analogous way to the effect of sodium dodecyl sulfate (SDS) in gel electrophoresis.

Use of denaturing conditions does generate issues. The viscosity created by the some agents (e.g. 6 M urea) particularly at lower temperatures may restrict flow rates due to increased back pressures. The separation range of the matrix is reduced as the hydrodynamic volume increases on denaturation. Whilst this may be advantageous in certain conditions the fractionation range quoted by the manufacturer for a standard aqueous mobile phase may not apply and unless specifically quoted by the supplier may need to be determined empirically. On denaturation multi-subunit proteins will be dissociated with the potential for erroneous molecular weight estimations and irretrievable loss of biological activity. Denaturing agents may also affect the monitoring of the chromatography due to interference in the far UV range. The use of denaturing agents may effect the stability of the column matrix or other components of the system such as the pumps.

Detergents are often added to the mobile phase when dealing with hydrophobic proteins such as membrane proteins. Most commercially available matrices are stable to the presence of commonly used detergents close to their critical micellar concentration. Some, like SDS are denaturing and act to disrupt lipid–protein interaction, often leaving proteins denatured with loss of biological function. Milder surfactants such as Triton X-100 can be used to leave some lipid association intact and in certain cases even retain biological function. As proteins partition into the micelle the mobility of the protein will increase due to the increased apparent molecular weight. Mobility may also be affected by the introduction of charged groups via association with ionic surfactants. Mobility may be altered by the micelle effect of salts and lipids which can induce mixed micelle formation. Surfactants such as Triton X-100 have strong absorbance at 280 nm which interferes with the monitoring of the chromatography. For this reason detergents such as reduced Triton X-100, Lubrol PX, and Brij 35 have been popular for chromatography. While non-ionic detergents are generally used when further purification steps are required such as ion exchange chromatography, SDS can be used if techniques such as reverse-phase chromatography are to be employed, although some effect on resolution in the subsequent step can be expected. As detergents such as SDS are difficult to remove from a column, manufacturers recommend that a column used with such additives in the mobile phase is dedicated.

2.2.3 Organic phases

Size exclusion of proteins in organic phases (sometimes called gel permeation) is sometimes used for membrane protein separations. The agarose and dextran based matrices are not suitable for separations with organic solvents. Many matrices have been designed for the separation of organic polymers in organic phases and these can often be applied to biopolymers such as proteins. For separation in aqueous organic mixed solvents polyacrylamide matrices are suitable. For totally organic systems synthetic polymer matrices, silica, and hydroxypropylated dextran (Sephadex LH20 and Sephadex LH60) are suitable.

A commonly used organic phase is FACE which is a mix of formic acid, acetic acid, chloroform, and ethanol in a ratio of 1:1:2:1 (19). The equipment must be compatible with the solvent system e.g. glass or Teflon. In general, separations in organic phases promote denaturation of the protein and hence an increase in hydrodynamic volume. This leads to a fractionation range lower and narrower in molecular weight than in aqueous phases.

2.3 Non-ideal size exclusion

Protein–matrix interaction is a common cause of protein loss or poor resolution during size exclusion. This may be due to complete retention on the column or retardation sufficient for the material to elute in an extremely broad dilute fraction so evading detection above the baseline of the buffer system. Many matrices retain a residual charge due, for example, to sulfate groups in agarose or carboxylic acid residues in dextran. Silica and glass require derivatization or coating to prevent adsorption. The silanol groups on the surface of silica are coated with a hydrophilic monolayer, usually glyceryl propylsilane. Nevertheless some cationic effects are still noted (15). With use hydrolysis of matrices at pH extremes can also introduce charged groups.

The ionic strength of the buffer should be kept at 0.15–2.0 M to avoid electrostatic or van der Waals interactions which can lead to non-ideal size exclusion (47, 48). Conversely, cross-linking agents such as those used in polyacrylamide may reduce the hydrophilicity of the matrix leading to the retention of some small proteins particularly those rich in aromatic amino acid residues. These interactions have been exploited effectively in some cases to enhance purification, however, they can often be negated by inclusion of organic modifiers such as acetonitrile and isopropanol in the mobile phase. Lectins can selectively bind to glucosidic functions of polysaccharide-based matrices.

If low ionic strength is necessary, then the risk of ionic interaction can be reduced by manipulating the charge on the protein via the pH of the buffer. This is best achieved by keeping the mobile phase above or below the pI of the protein as appropriate. High pI proteins may not chromatograph even in high ionic strength buffers. Very basic protein such as cytochrome c (pI ~ 10) and lysozyme (pI ~ 11) can elute after the V_t due to ionic interaction with the matrices. Recently, in our laboratory, a peptide of pI ~ 12.5 was seen to bind to agarose, dextran, and silica based SEC matrices. Elution was only facilitated by the use of ionic strengths equivalent to 5 M NaCl. Conversely acidic proteins may display ionic exclusion and elute early.

3 Methods

3.1 Flow rates

In chromatography, flow rates should be standardized for columns of different dimensions by quoting linear flow rate (cm/h). This is defined as the volumetric flow rate (ml/h) per unit cross-sectional area (cm^2) of a given column. As the principle of size exclusion is based on partitioning, success of the technique is particularly susceptible to variations in flow rates. Conventional low pressure size exclusion matrices tend to operate at linear flow rates of 5–15 cm/h. HPLC columns are run at 300–400 cm/h. Too high a flow rate leads to incomplete partitioning and band spreading. Conversely, very low flow rates may lead to diffusion and band spreading.

3.2 Preparation and packing

HPLC columns require specialized equipment for packing and are therefore supplied pre-packed by the manufacturer. Although experienced operators can successfully adjust and repair HPLC packings, for reasons of column performance and safety, it is generally advisable to avoid such operations. Low pressure SEC matrices can be purchased as pre-packed columns. They are also available as loose media, either pre-swollen or as dry powder.

If the gel is supplied as a dry powder, e.g. Sephadex (Pharmacia), it should be swollen in excess mobile phase as directed by the manufacturer. The conversion of dry weight to swollen gel volume is dependent upon the type of gel and the fractionation range. Sephadex G-10 swells to 2–3 ml/g whereas Sephadex G-200 swells to 20–45 ml/g. Swelling time is dependent upon temperature, Sephadex G-200 swells fully in 72 h at ambient temperature but within 5 h at 90°C. Preswollen gels are often supplied in a bacteriostat (e.g. 20% (v/v) ethanol). Occasionally a surfactant is included to give a more even slurry. Good column packing is an essential prerequisite for efficient resolution in SEC. The gel should be equilibrated and packed at the final operating temperature. The column can be packed under gravity although a more efficient method is to use a pump to push buffer through the packing matrix. The flow rate during packing should be approximately 50% higher than the operating flow rate (e.g. 15 cm/h for a 10 cm/h final flow rate).

Protocol 1

Preparation and packing of a low pressure size exclusion column

Equipment and reagents

- Empty chromatography column of appropriate dimensions
- Suitable mobile phase (e.g. PBS)
- Size exclusion matrix of appropriate fractionation range

A Swelling dry gel

1 Weigh out sufficient dry gel for the correct volume of matrices desired.[a]

2 Swell the gel in excess mobile phase. Avoid magnetic stirrers as these may damage the bead integrity.[b]

B Washing and degassing the gel

1 Pour the swollen gel into a sintered glass funnel and wash under low vacuum. If gel is supplied pre-swollen, resuspend the gel prior to decanting. Draw off the mobile phase under low vacuum.[c, d]

2 Wash the gel with five to ten column volumes of the final operating mobile phase and resuspend the gel by gentle agitation using a glass rod to a slurry of 50–75% (v/v).[e]

3 Transfer the equilibrated gel to a Buchner flask and allow to settle.

4 Remove fines from the top of the buffer layer by careful suction using a pipette.

5 Resuspend the gel to a 50–75% slurry and degas under low vacuum to remove any entrapped air.

C Packing and equilibrating the column

1 Support the column vertically in a retort stand, with the bottom flow adaptor in place according to the manufacturer's instructions. Where available, connect a packing reservoir to the top of the column.[f]

Protocol 1 continued

2 Carefully add a little degassed mobile phase in to the column to a level of 2 cm above the bottom end piece and allow to displace air and cap the outlet.

3 Apply the gel slurry to the column avoiding the inclusion of air bubbles. Pour the gel in one continuous manipulation.

4 Allow the gel to settle to within 20% of the packed volume.

5 Attach a pump, open the outlet, and apply the mobile phase at a flow rate 50% higher than the final operating flow rate or as directed by the manufacturer.[g]

6 Once a clear layer of buffer is established above the gel and the level of the gel remains constant, stop the flow, remove the packing reservoir, and apply the upper flow adaptor without disturbance to the top of the gel.

7 Continue the flow at 50% above the final operating flow rate for a minimum of two column volumes.

8 If a layer of buffer appears between the gel and the top flow adaptor carefully lower the adaptor onto the gel, avoiding disturbance to the gel.

9 Run the column at the packing flow rate until the level of the gel remains constant.

10 The gel can be either used or stored in the appropriate mobile phase with a preservative.

[a] The conversion of dry matrix weight to swollen gel volume differs from support to support. Consult the manufacturer's instructions for details.

[b] The rate of swelling is temperature dependent.

[c] The gel should never be allowed to dry out.

[d] Pre-swollen gels such as dextran and agarose have a sedimented volume approximately 15% greater than the packed column volume.

[e] It may be necessary to wash the gel first with deionized water if the storage buffer and mobile phase are not miscible.

[f] Avoid direct sunlight and sources of heat.

[g] Always pack under positive pressure. Do not draw buffer from the base of the column.

3.3 Equilibration

The packed column should be equilibrated by passing the final buffer through the column at an appropriate flow rate for at least two column volumes. Many pre-packed columns (e.g. HPLC columns) are supplied in preservatives such as ethanol. When changing the mobile phase it is prudent to use a flow rate lower (e.g. 50%) than the final operating flow rate. It is advisable to consider the miscibility of mobile phases prior to re-equilibration. Avoid switching directly from organic phases such as ethanol into mobile phases containing a high salt content and vice versa. If in doubt, it is advisable to include a transition phase such as distilled water to prevent salting-out of buffer components, etc. The pump should always be connected so as to pump the eluent on to the column under positive pressure. Drawing buffer through the column under negative pressure may lead to bubble formation. The effluent of the column should be sampled and tested for pH and conductivity in order to establish equilibration in the desired buffer.

3.4 Sample preparation and application

In many cases, a sample will require specific preparation prior to loading onto a SEC matrix. The optimum load of size exclusion columns is restricted to < 5% (v/v) (typically 2%) of the column volume in order to maximize resolution. The actual load volume limit will dependent upon column performance and resolution required. Whilst sample pH and ionic strength should not affect resolution, these parameters may need addressing with regard to sample stability. SEC columns are often loaded at relatively high concentrations of protein such as 2–20 mg/ml. The concentration is limited by solubility of the protein and the potential for increased viscosity which begins to have a detrimental effect on resolution. This becomes evident around 50 mg/ml. It is important to remove any insoluble matter prior to loading by either centrifugation or filtration.

3.5 Evaluation and calibration

SEC is a relative separation technique requiring calibration. Calibration is obtained by use of standard proteins and plotting V_e against log molecular weight. Successful calibration requires precise flow rates. The resultant plot gives a sigmoidal curve approaching linearity in the effective separation range of the gel (*Figure 2*). As a result of relative differences in hydrodynamic volume of the molecule it is critical to use the appropriate standards for size exclusion where proteins have a similar shape. Commonly used molecular weight standards are given in *Table 3*. Use of denaturing buffers (Section 2.2.2) and organic phases (Section 2.2.3) will affect the hydrodynamic volume of proteins and hence the calibration of matrices.

Table 3 Molecular weight standards for size exclusion chromatography

Standard	Molecular weight	Standard	Molecular weight
Vitamin B12	1350	Ovalbumin	43 000
Glucagon	3550	Bovine serum albumin	67 000
Aprotinin	6500	Alcohol dehydrogenase	150 000
Ribonuclease A	13 700	Bovine gamma globulin	160 000
Myoglobin	17 000	β-Amylase	200 000
Chymotrypsinogen A	25 000	Apoferritin	443 000
Carbonic anhydrase	29 000	Thyroglobulin	669 000
β-Lactoglobulin	37 000	Blue dextran 2000	~ 2 000 000

Protocol 2

Evaluation and calibration of a size exclusion column

Equipment and reagents

- Packed size exclusion chromatography column
- Suitable mobile phase (e.g. PBS)
- Solute for plate testing, e.g. 1% (v/v) acetone in mobile phase
- Standard proteins for calibration

A Equilibration

1 Equilibrate the system in the mobile phase with the column off-line in order to prime the system and displace any air.

2 Connect the column in-line ensuring that the direction of flow is appropriate or as specified by the column manufacturer.

3 Equilibrate the column in the desired mobile phase at a flow rate 50% lower than the final operating flow rate.[a,b]

4 After a minimum of two column volumes test for equilibration by checking the pH, and if possible the conductivity, of the effluent.[c]

5 If necessary continue to equilibrate the column until the column effluent is within 0.1 pH units of the mobile phase.

6 Adjust the pumps to the final operating flow rate, check the flow rate, and equilibrate for a further two column volumes.

B Evaluation and calibration

1 Where visible, e.g. low pressure columns, check the integrity of the gel bed. If necessary adjust the flow adaptors to eliminate any mobile phase at the top of the matrix.[c] Discard and re-pack a column which shows signs of gel cracking/channelling.

2 Plate test the column by injecting 5% of the column volume of 1% (v/v) acetone. Calculate the value of N and H as described in Section 1.1.2.[d]

3 Inject 5% of the column volume of a standard mix of proteins of the appropriate molecular weight range. The calibration can be made by plotting log mol. wt. against retention (V_e). Ensure that the separation is linear in the region required for target protein separation.[d]

[a] It may be necessary to wash the column first with deionized water if the storage buffer and mobile phase are not miscible.

[b] Always ensure that the column pressure does not exceed the maximum for the matrix.

[c] It is always prudent, where possible, to also check for conductivity.

[d] Where the load volume is expected to be less than 5%, calibration standards and test solute should be reduced in volume accordingly.

3.6 Separation of proteins

As described in Section 1.1.3 matrix selection, mobile phase, flow rate, column length, etc. are important factors influencing resolution. Although general principles exist for designing optimum separations in SEC, when sample availability is not limiting, separation can best be optimized by a test run. In addition to enabling alteration of parameters to improve resolution, a test separation enables adjustments to be made such as suboptimal increases in flow rate or sample volume where resolution attainable is in excess of that required. Due to the limitation on loading it is often wise to consider the suitability of the method for scale up when optimizing the operating parameters (*Figure 4*).

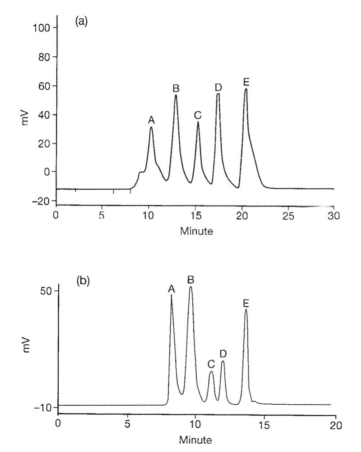

Figure 4 Size exclusion of standard proteins. (a) Protein separation on a Superdex 200 column. Size exclusion chromatography of five proteins on a Superdex 200 Prep Grade column (Pharmacia) (1.6 × 75 cm). (A) Thyroglobulin 670 000; (B) gamma globulin 150 000; (C) ovalbumin 44 000; (D) myoglobin 17 000; (E) vitamin B_{12} 1350. (b) Peptides on Superdex Peptide column. Size exclusion chromatography of five peptides on a Superdex peptide analytical column (Pharmacia) (1 × 30 cm). (A) Cytochrome *c* 12 500; (B) aprotinin 6500; (C) angiotensin II 1046; (D) GlyAlaTyr 309; (E) GlyLeu 188.

Protocol 3

Separation of proteins by size exclusion chromatography

Equipment and reagents

- Packed chromatography column of appropriate dimensions
- Suitable mobile phase (e.g. PBS)
- Standard low or high pressure chromatography equipment with manual or electronically operated injector

Method

1 Connect the column to chromatography system and equilibrate/calibrate the matrix as described in *Protocol 2*.

2 Set column flow rate to operational level.[a]

3 Check column integrity for cracks or channelling and equipment for any leaks.

4 Prepare the fraction collector to collect the appropriate number of fractions.

5 Monitor the detector output to ensure a stable baseline is achieved prior to loading.

6 Remove any particulate matter by centrifugation immediately prior to loading.[b]

7 If using an injection loop, displace any air formed by injecting mobile phase.[c]

8 Apply the sample at a flow rate 50% that of the operational flow rate.

9 Reposition the valve to the load position and elute proteins at the operational flow rate.

10 Collect fractions for further resolution/assay.

11 The molecular weight of proteins of a given V_e can be calculated by the method described in Protocol 2.

[a] Flow rate can be checked by timed collection into a graduated measuring cylinder.

[b] For small samples of < 1.5 ml a normal microcentrifuge (13 000 r.p.m.) is usually adequate although use of 0.45 or 0.2 μm filtration may be prudent.

[c] The sample loop should ideally have a volume equal to the volume to be loaded.

3.7 Column cleaning and storage

Size exclusion matrices can be cleaned *in situ* or as loose gel in a sintered glass funnel. Cleaning is particularly recommended prior to storage and when increased back pressure across the column is observed. Suppliers usually offer specific guidelines for cleaning gels. Common general cleaning agents include non-ionic detergents (e.g. 1% (v/v) Triton X-100) for lipids and 0.2–0.5 M NaOH for proteins and pyrogens (not recommended for silica based matrices). In extreme circumstances contaminating protein can be removed by use of enzymic digestion (pepsin for proteins and nucleases for RNA and DNA). The gel should be stored in a buffer with antimicrobial activity such as 20% (v/v) ethanol or 0.02–0.05% (w/v) sodium azide. NaOH is a suitable storage agent which combines effective solubilizing activity with prevention of endotoxin formation. It may, however, lead to chemical breakdown of certain matrices. Many matrices are stable to autoclaving under defined conditions.

3.8 Trouble shooting

In size exclusion the macromolecules are not physically retained, unlike adsorption techniques, therefore the protein will elute in a defined volume between V_0 and V_t. If the protein elutes before the void volume ($V_e < V_0$) this suggests channelling through the column due to improper packing or operation of the column or ionic exclusion. If the protein elutes after the total volume ($V_e > V_t$) then some interaction must have occurred between the matrix and the protein of interest.

In addition to determination of the number of theoretical plates (N), the performance of the column can be assessed qualitatively by the shape of the eluted peaks (*Figure 5*). The theoretically ideal peak is sharp and triangular with an axis of symmetry around the apex. Deviations from this are seen in practice. Some peak shapes are diagnostic of particular problems which lead to broadening and poor resolution. If the downslope of the peak is significantly shallow it is possible that the concentration of the load was too high or the material has disturbed the equilibrium between the mobile and stationary phases. If the downslope tends towards being symmetrical initially but then tails it is common to assume that there is a poorly resolved component, although it may be suspected that interaction with the matrix is taking place. A shallow upslope may represent insolubility of the loaded material. A valley between two closely eluting peaks may suggest poor resolution, but can also be the result of a faulty sample injection.

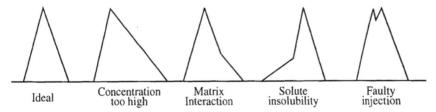

| Ideal | Concentration too high | Matrix Interaction | Solute insolubility | Faulty injection |

Figure 5 Diagnosis of column performance.

References

1. Lathe, G. H. and Rutheven, C. R. (1955). *J. Biochem.*, **60**, 34.
2. Porath, J. and Flodin, P. (1959). *Nature*, **183**, 1657.
3. Moore, J. C. (1964). *J. Polymer Sci.*, **A2**, 835.
4. Ettre, L. S. (1993). *Pure Appl. Chem.*, **65**, 819.
5. Laurent, T. C. (1993). *J. Chromatogr.*, **633**, 1.
6. Burnouf, T. (1995). *J. Chromatogr.*, **664**, 3.
7. Katakam, M. and Banga, A. K. (1994). *J. Pharm. Sci. Technol.*, **49**, 160.

8. Sebille, B. (1990). *Chromatogr. Sci.*, **51**, 585.

9. Uversky, V. N. (1993). *Biochemistry*, **32**(48), 13288.

10. Uversky, V. N. (1994). *Int. J. Biochromatog.*, **1**(2), 103.

11. Ackland, C. E., Berendt, W. G., Freeza, J. E., Landgraf, B. E., Pritchard, K. W., and Ciardelli, T. C. (1991). *J. Chromatogr.*, **340**, 187.

12. Determann, H. (1969). In *Advances in chromatography* (ed. J. C. Giddings and R. A. Keller), Vol. 8, p. 35. Marcel Dekker Inc., New York.

13. Basedow, A. M., Ehert, K. H., Ederer, J. H., and Fosshag, E. (1980). *J. Chromatogr.*, **192**, 259.

14. Squire, P. G. (1985). In *Methods in enzymology* (ed. C. H. W. Hus and S. N. Timasheff), Vol. 117, p. 142.

15. Himmel, M. E., Baker, J. O., and Mitchell, D. J. (1995). *Chromatogr. Sci. Ser.*, **69**, 409.

16. Mant, C. T. and Hodges, R. S. (1989). *J. Liq. Chromatogr.*, **12**, 139.

17. Stellwagen, E. (1990). In *Methods in enzymology* (ed. M. P. Deutscher), Vol. 182, p. 317.

18. Berglof, J. H., Eriksson, S., and Andersson, I. (1987). *Dev. Biol. Standard.*, **67**, 25.

19. Findlay, J. B. C. (1990). In *Protein purification applications* (ed. E. L. V. Harris and S. Angal), p. 59. IRL Press, Oxford.

20. Hjemeland, L. M. (1990). In *Methods in enzymology* (ed. M. P. Deutscher), Vol. 182, p. 277.

21. Lindqvist, L. O. and Williams, K. W. (1973). *Dairy Ind. Int.*, **38**, 459.

22. Friedli, H. and Kistler, P. (1972). *Chimia*, **26**, 25.

23. Freitag, R. and Horvath, C. (1996). *Adv. Biochem. Eng. Biotechnol.*, **53**, 17.

24. Ladisch, M. R. and Kohlmann, K. L (1992). *Biotechnol. Prog.*, **8**, 469.

25. Cole, E. S., Lee, K., Lauzere, K., Kelton, C., Chappel, S., Weintraub, B., *et al.* (1993). *Biotechnology*, **11**, 1014.

26. Pohl, T. (1990). In *Methods in enzymology* (ed. M. P. Deutscher), Vol. 182, p. 69.

27. Laue, D. M. and Rhodes, D. G. (1990). In *Methods in enzymology* (ed. M. P. Deutscher), Vol. 182, p. 566.

28. Soreghan, B., Kosmoski, J., and Glabe, C. (1994). *J. Biol. Chem.*, **269**, 28551.

29. Hagel, L. (1993). *J. Chromatogr.*, **648**, 19.

30. Garnick, R. L, Ross, M. J., and DuMee, C. P. (1988). In *Encyclopedia of pharmaceutical technology* (ed. J. Swarbrick and J. C. Boylan), Vol. 1, p. 253. Marcel Dekker Inc., New York.

31. United States Food and Drugs Adminisration (1994). FDA points to consider in the manufacturing of monoclonal antibody products for human use.

32. Huang, J.-X. and Guiochon, G. (1989). *J. Chromatogr.*, **492**, 431.

33. Feisier, H. and Gooding, K. M. (1991). *J. Chromatogr.*, **544**, 125.

34. Schnabel, R. and Langer, P. (1991). *J. Chromatogr.*, **544**, 137.

35. Berman, R. J., Renn, C. N., and Johnson, E. L. (1994). *Anal. Chem.*, **66**, 1R.

36. Johansson, H., Ostling, M., Sofer, G., Wahlstrom, H., and Low, D. (1988). *Adv. Biotechnol. Processes*, **8**, 127–57.

37. Sofer, G. K. and Nystrom, L. E. (1989). *Process scale chromatography: a practical guide.* Academic Press.

38. Curling, J. M. and Cooney, J. M. (1982). *J. Parental Sci. Technol.*, **36**, 59.

39. Barth, H. G., Boyes, B. E., and Jackson, C. (1994). *Anal. Chem.*, **66**, 595.

40. Stuting, H. H. and Krull, I. S. (1990). *Anal. Chem.*, **62**, 2107.

41. Huber, A. (1991). *Biochem. Soc. Trans.*, **19**, 505.

42. Li, Y. T., Hsieh, Y. L., Henion, J. D., Senko, M. W., McLafferty, F. W., and Ganern, B. (1993). *J. Am. Chem. Soc.*, **115**, 8409.

43. Nylander, I., Tan-No, K., Winter, A., and Silberring, J. (1995). *Life Sci.*, **57**, 123.

44. Hensel, J., Hintz, M., Karas, M., Linder, D., Stahl, B., and Geyer, R. (1995). *Eur. J. Biochem.*, **232**, 373.

45. Glocker, M. O., Arbogast, B., Milley, R., Cowgill, C., and Deinser, M. L. (1994). *Proc. Natl. Acad. Sci. USA*, **91**, 5868.
46. Hagel, L. (1989). In *Protein purification* (ed. J.-C. Jansen and L. Ryden), p. 63. VCH Publishers Inc.
47. Kopaciewicz, W. and Regnier, F. E. (1982). *Anal. Biochem.*, **126**, 8.
48. Dubin, P. L., Edwards, S. L., Mehta, M. S., and Tomalia, D. (1993). *J. Chromatogr.*, **635**, 51.

Chapter 9

Purification by exploitation of activity

S. Angal
Baxter Highland Division, Glendale, California.

P. D. G. Dean
Cambridge Life Sciences plc, Cambridge Science Park, Milton Road, Cambridge CB4 4GN, UK.

1 Introduction

Proteins carry out their biological functions through one or more binding activities and, consequently, contain binding sites for interaction with other biomolecules, called ligands. Ligands may be small molecules such as substrates for enzymes or larger molecules such as peptide hormones. The interaction of a binding site with a ligand is determined by the overall size and shape of the ligand as well as the number and distribution of complementary surfaces. These complementary surfaces may involve a combination of charged and hydrophobic moieties and exhibit other short-range molecular interactions such as hydrogen bonds. This binding activity of a protein, which is stereoselective and often of a high affinity, can be exploited for the purification of the protein in a technique commonly known as affinity chromatography.

The operation of affinity chromatography involves the following steps:

(a) Choice of an appropriate ligand.

(b) Immobilization of the ligand onto a support matrix.

(c) Contacting the protein mixture of interest with the matrix.

(d) Removal of non-specifically bound proteins.

(e) Elution of the protein of interest in a purified form.

At best, affinity chromatography is the most powerful technique for protein purification since its high selectivity can, in principle, allow purification of a single protein of low abundance from a crude mixture of proteins at higher concentrations. Secondly, if the affinity of the ligand for the protein is sufficiently high, the technique offers simultaneous concentration from a large volume. In practice, such single-step purifications are not common and successful affinity chromatography requires careful consideration of a number of parameters involved. The remainder of this chapter attempts to guide the experimenter in the selection and use of affinity adsorbents for protein purification. For more extensive information on this technique the reader is advised to consult the many excellent texts on this subject (1–7) as well as proceedings of symposia (8–11).

2 Design and preparation of affinity adsorbents

The construction of an affinity adsorbent for the purification of a particular protein involves three major factors:

(a) Choice of a suitable ligand.

(b) Selection of a support matrix and spacer.

(c) Attachment of the ligand to a support matrix.

The criteria for making these decisions are discussed in the following sections.

2.1 Choice of ligand

Suitable pairs of protein and ligand combinations for affinity chromatography are antigen–antibody, hormone–receptor, glycoprotein–lectin, or enzyme–substrate/cofactor/effector. In practice, the natural biological ligand may be very expensive or difficult to obtain and analogues or mimics (pseudo-ligands) are preferentially employed. The factors to consider when selecting a ligand for protein purification are as follows:

(a) **Specificity**. Ideally, the ligand should recognize only the protein to be purified (e.g. monospecific antibody). The choice may be evident from the known biological properties of the protein to be purified. If this is not possible a ligand known to recognize a group of proteins should be selected. Examples of ligands are given in *Table 1*. Group-specific ligands, by definition, allow the purification of related proteins or protein families without the need to invest a large amount of time and labour in the preparation of monospecific adsorbents. Their wide application also makes them popular with manufacturers of affinity adsorbents and many are, therefore, available in ready-to-use format. In combination with other increasingly high-resolution chromatographic techniques they offer the advantage of speed over tailor-made monospecific adsorbents.

(b) **Reversibility**. The ligand should form a reversible complex with the protein to be purified such that the complex is resistant to the composition of the feedstream and washing buffers but is easily dissociable without requiring denaturing conditions for elution.

(c) **Stability**. The ligand should be stable to the conditions to be used for immobilization (e.g. organic solvents in some cases) as well as the conditions of use. This includes resistance to proteolysis and to denaturation by eluents or cleaning agents.

(d) **Size**. The ligand should be large enough such that it contains several groups able to interact with the protein resulting in sufficient stereoselectivity and affinity. In addition, the ligand should contain a functional group which can be used for immobilization without significantly affecting the protein-binding characteristics. If the ligand is so small that the matrix backbone would interfere with access to protein molecules it is preferable to interpose a spacer molecule between the ligand and matrix. On the other hand, a ligand which is very large is likely to be more susceptible to denaturation or degradation and can cause increased non-specific binding through other parts of the molecule. The size of the ligand determines the amount which can be coupled to a matrix and its efficiency. This matter is discussed in more detail in Sections 3.3 and 3.4.

(e) **Affinity**. The interaction of a protein P and a ligand L can be described by the equation:

$$P + L \leftrightarrow PL$$

where the equilibrium dissociation constant, K_L, is defined by:

$$\frac{[P]\,[L]}{[PL]}$$

A simple rearrangement gives:

$$\frac{[L]}{K_L} = \frac{[PL]}{[P]}$$

which implies that for substantial adsorption of the protein from solution (e.g. $[PL] : [P] = 95 : 5$) the value of K_L must be about two orders of magnitude less than the concentration of immobilized ligand.

In practice, for a protein–ligand of M_r 10 000 immobilized at 10 mg/ml (i.e. 1 mM) a value for K_L less than 10^{-5} M would be needed for substantial adsorption of the desired protein from similar volumes of matrix and solution (operation in batch mode). For concentration from a tenfold volume the K_L should be less than 10^{-6} M. Smaller ligands can be immobilized at 10–100-fold higher concentrations and would, by analogy, be able to cope with a higher K_L. For higher ligand affinities ($K_L < 10^{-8}$ M) dissociation of the ligand–protein may prove difficult. As a general rule affinity techniques operate well between $K_L = 10^{-4}$ and 10^{-8} M. Batch techniques and low ligand concentrations may be successful at the lower K_L whereas columns and higher ligand concentrations will be required for $K_L = 10^{-4}$ M. These numbers can only be used as guidelines since the K_L values determined for free ligand may be different to those for the immobilized ligand and the proportion of immobilized ligand which is functional varies with ligand size as will be discussed later (Section 3.4).

2.2 Selection of the matrix

Table 2 shows examples of commercially-available affinity adsorbents. A chosen affinity ligand is immobilized to a solid support or matrix via one or more co-

Table 1 Examples of ligands suitable for the purification of proteins by affinity chromatography

Ligands[a]	For purification of
Protein A/G	Immunoglobulins from various species
Monoclonal antibodies	Antigens, protein A fusions
Antigens	Specific monoclonal/polyclonal antibodies
Steroids	Steroid receptors, steroid binding proteins
Fatty acids	Fatty acid-binding proteins, albumin
Nucleotides	Nucleic acid-binding proteins, nucleotide-requiring enzymes
Protease inhibitors	Proteases
Lysine/arginine	Plasmin, plasminogen activators
Polymixin B	Bacterial endotoxin C, removal of endotoxin, from protein
Phenylboronate	Glycated proteins, glycoproteins
Lectins	Glycoproteins
Sugars	Lectins, glycosidases
Phosphoric acid	Phosphatases
Biotin	Biotin-binding proteins, avidin, streptavidin
Avidin	Biotin-containing enzymes
Heparin	Coagulation factors, lipases, connective tissue, proteases, DNA polymerases
Gelatin	Fibronectin
Calmodulin	Calmodulin-binding enzymes
Triazine dyes and mimetic ligands	Dehydrogenases, kinases, polymerases, interferons, restriction enzymes

[a] All the above ligands are available in the immobilized form from a variety of suppliers.

valent bonds. The effectiveness of the immobilized ligand in purifications can be markedly dependent on the structure of the matrix (12). It is, therefore, important to consider some criteria for the selection of matrices:

(a) The matrix should have a high degree of porosity so that large proteins can have unhindered access to ligand immobilized on the interior portions of the lattice.

(b) The matrix should be chemically stable under the conditions used for activation and coupling as well as those used during operation and regeneration.

(c) The matrix should be physically rigid in order to allow good flow properties and able to withstand some mechanical agitation without breaking up.

(d) The matrix should withstand a reasonable range of pH and temperature.

(e) The matrix should be easily activatable for coupling of ligands at high density yet be otherwise inert to non-specific binding of proteins.

(f) Matrix properties should not be substantially altered on functionalization.

(g) The matrix should be uniform in structure, particularly, when functionalized so that ligand molecules can be homogeneously distributed.

Table 2 Examples of commercially available affinity absorbents

Supplier	Material	Specificity	Ligand
Bio-Rad	Affi-gel protein A	IgG	Protein A
	Affi-prep protein A	IgG	Protein A
	Affi-gel Blue	Albumin, nucleotide-requiring enzymes	Cibacron Blue
Pierce	Immobilized heparin	Blood proteins	Heparin
	Immobilized jacalin	IgA	Jacalin
	Immobilized pepstatin	Cathepsin	Pepstatin
BioSepra	Protein A hyper D	IgG	Protein A
	Heparin hyper D	AT111, FIX	Heparin
	Blue trisacryl	Albumin, interferon	Blue dye
Amersham Pharmacia	Lentil lectin Sepharose-4B	Mannose, glucose	Lentil haemagglutinin
	Protein G Sepharose	IgG	Protein G
	Chelating Sepharose	Proteins via metal ion complexes	Iminodiacetic acid
PerSeptive	Poros A and	IgG	Protein A and
	G Poros HE	Blood proteins	G Heparin
Prometic Biosciences/ACL	Mimetic	Wide range of proteins	Synthetic

For any particular application the importance of the above criteria may differ; for example, incompressibility would be more important at higher pressures or diffusibility would be more important for purification of very large proteins.

Comparative studies of matrices (13), generally with model proteins, have indicated that agarose is one of the better matrix materials. The novice is, therefore, advised to use agarose or cross-linked agarose unless otherwise indicated by the nature of his or her application. Systematic comparisons of matrices may be carried out at a later stage, preferably using similar ligand densities, in order to improve a process or circumvent a problem such as leakage.

A large number of matrices and ligands are now available. For examples, refer to *Tables 2* and *3*. The interested reader should obtain further information from the manufacturers.

2.3 Choice of spacer

A spacer molecule may be employed to distance the ligand from the matrix backbone in applications where the small size of the ligand excludes it from free access to protein molecules in the solvent (4, 5, 7). A spacer molecule may also be effective if the ligand is immobilized through a site near enough to the protein-binding surface to interfere with protein binding.

The length of the spacer arm is crucial and must be determined empirically. The number of methylene groups most often successful is six to eight. The spacer should be hydrophilic and not itself bind proteins, either because of its hydrophobicity or its charged groups.

Table 3 Examples of commonly available activated matrices

Activated matrix	Properties
Cyanogen bromide activated	Activation of polysaccharides, particularly agarose; highly popular; cyanogen bromide is toxic—use commercially activated agarose
Carbonyldiimidazole	Activation of polysaccharide matrices using carbonyldiimidazole to give reactive imidazole carbonate derivatives
Oxiranes	Bis-oxiranes react with amino- or hydroxyl-containing gels at alkaline pH; highly stable
Sulfonyl chloride	Reaction of organic sulfonyl halides with matrix hydroxyl groups and subsequent coupling with amino and thiol groups
Periodate	Provides stable links to proteins using periodate oxidation and ideal when ligand is sensitive to alkaline pH
Fluoromethyl pyridinium sulfonate	Activation of polysaccharide hydroxyls in presence or tertiary amine such as triethylamine in a polar solvent. Subsequent reaction with amino- or sulfydral-containing ligands
Glutaraldehyde	Activation by glutaraldehyde of amino or amide functions followed by coupling with primary amines

2.4 Activation and coupling chemistry

The detailed methodology of activation of support matrices is beyond the scope of this chapter and the reader is referred to a previous text in this series (1). A brief description of the major methods is given below to allow the reader to grasp the salient points of each method before selecting pre-activated matrices which are available commercially (*Table 3*).

2.4.1 Cyanogen bromide

Cyanogen bromide (CNBr) reacts with hydroxyls in agarose and other polysaccharide matrices to produce a reactive support which can subsequently be derivatized with spacer molecules or ligands containing primary amines (14). The N-substituted isourea formed on reaction with the primary amine is positively charged at physiological pH (pK_a ∼ 9.5) thus imparting anion exchange properties to the adsorbent. The isourea derivative is susceptible to nucleophilic attack (e.g. amine-containing buffers, proteins) and slow hydrolysis can occur at extremes of pH resulting in leakage of ligand into the medium. CNBr is extremely toxic (releases HCN on acidification) so commercially activated agarose is recommended. Even with these limitations, CNBr activation is one of the most widely used processes in the preparation of affinity adsorbents.

Unprotonated primary amines (except Tris) couple efficiently to CNBr-activated agarose, therefore the coupling pH should be greater than the pK_a of the ligand but less than 10 (e.g. pH 7–8 for aromatic amines, pH 10 for aliphatic amines). Borate or carbonate buffers are recommended. The coupling efficiency decreases at pH values above 9.5–10.0 which reflects the sharp decline in stability of the activated complex. CNBr-activated agarose is not stable at elevated temperatures and the coupling should be carried out at +4°C. For details of the coupling procedure see Section 4.2.

2.4.2 Bis-oxirane

Bis-oxiranes (bis-epoxides) react readily with both hydroxyl- or amino-containing gels at alkaline pH to yield derivatives which possess a long chain hydrophilic reactive oxirane. These in turn may be reacted with nucleophiles to prepare affinity adsorbents with lower numbers of hydrophilic and ionic groups than those obtained with CNBr activation. The procedure automatically introduces a long chain hydrophilic spacer arm which may be desirable in certain applications. Coupling reactivity is in the order SH > NH > OH and the efficiency is enhanced by raising the temperature to 40°C. Oxirane-coupled ligands are extremely stable.

2.4.3 Carbonyldiimidazole

Polysaccharide matrices may be activated using carbonylating agents such as N,N' carbonyldiimidazole (CDI) under anhydrous conditions to give reactive imidazole carbonate derivatives (15). These in turn will react with ligands containing primary amino groups at alkaline pH to give stable carbonate derivatives.

CDI is non-toxic, gives high levels of activation, and does not introduce ion exchange groups into the matrix. CDI-activated agarose (available from Pierce) is stable in anhydrous conditions (but not in aqueous solution) and supplied in acetone. Before coupling the acetone should be removed by quickly washing the gel on a sintered glass funnel with ice-cold water (not necessary if the ligand is insensitive to acetone). Coupling is recommended in 0.1 M borate buffer pH 8.5–10 at room temperature.

2.4.4 Sulfonyl chloride

Organic sulfonyl halides react with matrix hydroxyl groups to form sulfonyl esters which are themselves excellent leaving groups (16). Nucleophiles (e.g. amino and thiol groups of proteins) will readily displace the sulfonate enabling efficient and rapid coupling. Activation using p-toluene sulfonyl (tosyl) or tri-fluoroethyl sulfonyl (tresyl) chloride is carried out in organic solvents and the resulting matrix is stable (in 1 mM HCl or in the dry state) for extended periods. The reactivity of the sulfone ester is strongly influenced by the substituent attached to the sulfonate. Thus, tresyl displacement can occur at neutral pH and 4°C while tosyl-activated matrices need to be coupled in alkaline pH (9–10.5, bicarbonate buffers) at 40°C. An advantage of the tosyl-activated matrix is that tosyl displacement can be followed spectrophotometrically.

2.4.5 Periodate

Vicinal diol groups of polysaccharide matrices may be oxidized by the use of sodium m-periodate (NaIO$_4$) to generate aldehyde functions. NaIO$_4$ is very soluble in water and can, therefore, be easily removed from the activated gel by washing. The activated gel is stable at 4°C for several days. The aldehyde groups react with primary amines at pH 4–6 to form a Schiff's base which can be stabilized by reduction with sodium borohydride (or cyanoborohydride). Details of activation and coupling follow:

(a) Mix the gel with an equal volume of 0.2 M NaIO$_4$ and place in a tightly closed polythene bottle.

(b) Allow to mix gently for about 2 h at room temperature.

(c) Wash the gel thoroughly on a sintered glass funnel with distilled water and then with 0.5 M phosphate buffer pH 6.0.

(d) Add the gel to an equal volume of 0.5 M phosphate buffer pH 6.0 containing about 25 mM of the desired amine ligand and 0.5 mM sodium cyanoborohydride.

(e) Allow to mix gently for three days at room temperature.

(f) Wash the gel exhaustively and reduce any remaining aldehydes by incubation with 1 M sodium borohydride for 15 h at 4°C.

(g) Wash the gel extensively in distilled water and then in storage buffer.

Periodate oxidation represents a convenient alternative to other methods, particularly, where ligands are sensitive to alkaline pH. The reagents are non-toxic and the alkylamine product is stable.

2.4.6 Fluoromethyl pyridinium sulfonate

Facile activation of polysaccharide hydroxyls can be carried out using 2-fluoro-1-methylpyridinium toluene-4-sulfonate (FMP) in polar organic solvents (17). The reaction is carried out in the presence of a tertiary amine such as triethylamine and is completed in 10 min at room temperature. The resulting activated groups (2-alkoxy-1-methylpyridinium salts) can react readily (pH 8–9, aqueous or polar organic) with amino- or sulfydryl-containing ligands. Both activation and coupling can be followed spectrophotometrically since 1-methyl-2-pyridine, the leaving group, absorbs strongly at 297 nm. FMP is non-toxic, inexpensive, and claimed to form stable, non-ionic linkages (thioether or secondary amine) at high degrees of substitution.

2.4.7 Glutaraldehyde

Activation of amino or amide functions (e.g. polyacrylamide or agarose with amine spacers) can be carried out using 25% aqueous (fresh) solutions of glutaraldehyde (0.5 M phosphate buffer pH 7.5, 20–40°C, overnight). The gel should be washed to remove the aldehyde and then coupling of primary amines carried out in the same buffer but at 4°C. This method is simple, inexpensive, and suitable for ligands sensitive to alkaline pH. The linkage is stable and simultaneously introduces a spacer arm. Glutaraldehyde can polymerize on storage and is moderately toxic.

2.4.8 Cross-linking techniques

Nucleophilic ligands (e.g. containing amino groups) can be coupled to carboxylated matrices using carbodiimides which promote condensation to generate peptide bonds (19).

Dicyclohexyl carbodiimide is insoluble in water and must be used in organic solvents such as dioxane, dimethyl sulfoxide, 80% (v/v) aqueous pyridine, or acetonitrile. A problem encountered in these condensations is the removal of insoluble ureas and other by-products which must be washed from the beads by washing in organic solvents.

Water soluble carbodiimides are more convenient since their corresponding ureas are much more soluble. An example is described below:

(a) Prepare carboxylated matrix (e.g. CH-Sepharose) by washing extensively in water and adjusting to pH 4.0–4.5. Do not use amino-, hydroxyl- or phosphate-containing buffers throughout the reaction.

(b) Freshly prepare a solution of 1-ethyl-3-(3-dimethyl aminopropyl)-carbodiimide (40 mg in 10 ml of buffer at pH 4.5). Add to 4 ml of matrix.

(c) Add an amino-containing ligand (10 × concentration of spacer which is usually 10–20 μmol/ml) and mix by end-over-end rotation at room temperature for at least 1 h but up to 2 h.

(d) Terminate the reaction by filtering the matrix and wash out the unreacted carbodiimide and by-products.

2.4.9 Triazine dyes

Triazine dyes are 'reactive' dyes containing mono or dichlorotriazinyl groups which may be coupled directly to hydroxyl-containing matrices using the following procedure:

(a) Wash the matrix (e.g. agarose) with water, weigh out the moist gel (20 g), and suspend it in 50 ml of water.

(b) Weigh 200 mg of dye (available from Sigma) and add 20 ml of water.

(c) Mix the matrix and dye solution thoroughly.

(d) Add 10 ml of NaCl (20%, w/v) solution and incubate at room temperature for 30 min.

(e) Add 20 ml of Na_2CO_3 (5%, w/v) and incubate in a 45 °C water-bath (dichloro-triazines couple in 1 h whereas monochlorotriazines require 40 h).

(f) Wash the coupled matrices with warm water, 6 M urea, and 1% (w/v) Na_2CO_3 until the washings are colourless. Blocking is not required. Store in 0.1% (w/v) Na_2CO_3 containing 0.02% (w/v) sodium azide.

Refer to Section 3.5 for information on the choice of dyes and their use.

2.4.10 Blocking reactive groups

Many commercial pre-activated supports have reactive group concentrations in the range 5–25 μmol/ml while coupled ligand concentrations are in the range 1–10 μmol/ml. Thus, a number of residual activated groups will remain on gels after coupling. These groups may be blocked by the addition of low molecular weight compounds. The most common blocking agent is 1 M ethanolamine adjusted to pH 8.0 and incubated with the matrix at room temperature for 1 h. If other small molecules are used (e.g. glycine) their effect on the level of non-specific binding should be examined. At the end of the blocking period the adsorbents should be washed using the most stringent conditions recommended by the manufacturer before storage in the presence of preservatives (e.g. 0.02% sodium azide or methiolate). Other conditions of storage will depend on the ligand and matrix.

2.5 Estimation of ligand concentration

It is essential to determine the success of a ligand immobilization procedure at the time of coupling. This can be done using one of the following methods.

2.5.1 Difference analysis

Measure the amount of ligand added to the coupling mixture and that recovered after the washing procedures. The method used for measurement will depend on the nature of the ligand (e.g. spectrophotometry, immunoassay, activity assay).

2.5.2 Direct measurements

(a) Suspend derivatized matrix in 50% (v/v) glycerol, mix thoroughly, and measure absorbance against a similarly treated suspension of underivatized matrix. This method is suitable for ligands absorbing at wavelengths different from that at which the matrix absorbs.

(b) Alternatively, derivatized gels may be assayed for protein using the Lowry's method. Again, it is necessary to use the underivatized gel as a control.

(c) If the immobilized ligand contains a unique group (e.g. phosphate) direct elemental analysis is most appropriate.

(d) If the ligand contains a chemically reactive group (e.g. thiol) appropriate measurements can be made after reaction.

2.5.3 Radioactive ligand

Radioactive ligand may be incorporated into the coupling reaction to give the most sensitive determination of total immobilized ligand.

2.5.4 Hydrolysis

This method is suitable only if the hydrolysed ligand can be measured in the presence of hydrolysis products of the matrix. Agarose can be solubilized by heating to 90°C in 50% (v/v) acetic acid or 0.5 M HCl for about 1 h. Alternatively, more vigorous hydrolysis (e.g. 6 M HCl at 110°C *in vacuo* for 24 h) will liberate amino acids from agarose derivatized with protein ligands. Cross-linked agaroses are more resistant to hydrolysis. Immobilized ligands may be hydrolysed using suitable enzymes (e.g. alkaline phosphatase for nucleotides and pronase for proteins). However, release of ligands reflects their accessibility to the digesting enzymes and the method is likely to underestimate ligand concentration.

3 Use of affinity adsorbents

3.1 Initial questions

Having prepared or obtained an affinity absorbent the first question is whether it will bind and purify the protein of interest. This question can be answered after the following experiment.

(a) Using 1 ml of the adsorbent prepare a suitable column (1–2 ml capacity plastic columns with 10 ml reservoirs are widely available).

(b) Equilibrate the column in a buffer chosen to encourage optimum binding of ligand and protein. Some guidance is provided in a later section, otherwise a probable set of conditions can be chosen for the known properties of the ligand and protein pair.

(c) Prepare the crude protein extract, preferably in or dialysed against the equilibration buffer. This should be as concentrated as possible.

(d) Apply a small aliquot of the extract to the column. It should contain about ten times the amount detectable in a specific assay for the protein to be purified and should not exceed the probable capacity of the column. Assume that 1% of the immobilized ligand is likely to be functional where ligands are small and immobilized at high concentrations. For protein ligands the capacity may be 10% of the immobilized ligand.

(e) Wash the column with 10 ml of equilibration buffer and collect 1 ml fractions.

(f) Elute bound proteins with 5 ml of equilibration buffer containing an eluent. If a possible eluent is not known try 1 M NaCl (5 ml) followed by 0.1 M glycine–HCl pH 2.5.

(g) Analyse all collected fractions for total protein content and for the protein of interest. Calculate per cent purity in each fraction. Four possible results may be obtained.

  (i) The protein of interest is found in fractions 1–3 of the wash together with most of the contaminating protein. It, therefore, has no affinity for the adsorbent and the choice of the adsorbent is in question.

  (ii) The protein binds and is eluted by one of the eluents with improved purity. This is the ideal situation and a larger scale purification could follow.

  (iii) The protein of interest is found in fractions 4–10 (i.e. it is retarded by the adsorbent) with improved purity.

  (iv) The protein of interest is not found in any of the fractions implying that it is still bound to the column or destroyed during the course of the experiment.

In case of situation (iii) or (iv) it is necessary to manipulate the operating conditions as described in the next section.

3.2 Selection of conditions for operation

These studies should also be carried out on the small scale described above.

3.2.1 Choice of adsorption conditions

The buffer chosen to effect adsorption should reflect the conditions required to achieve a strong complex of the ligand with the protein to be purified. As indicated in Section 1 the binding interaction may be mediated by a variety of molecular forces. If the interaction is thought to be predominantly hydrophobic an increase in the ionic strength and/or pH will improve adsorption. An example of such an improvement is the binding of protein A to murine immuno-globulins. Other interactions may be reinforced by the addition of divalent metal ions or specific factors able to preserve a particular protein conformation. Some knowledge of the ligand–protein interaction may suggest other ways of improvement. (For example, lower temperatures weaken hydrophobic interaction.) Some trial and error experiments are required if nothing is known about the interaction.

A second method for improving the affinity of the adsorbent for a protein is to increase the concentration of the immobilized ligand by increasing the amount and concentration of ligand used at the coupling stage. It may be necessary to start with a matrix containing a higher level of activated groups.

Stronger binding can, sometimes, be promoted by allowing longer incubation times (20). This effect may be due to secondary binding interactions. Consider operation in batch mode.

Further improvements can be made on scaling-up; for example, longer column length, slower flow rate, and lower sample volumes will all allow for increased retardation.

3.2.2 Choice of washing conditions

If the ligand–protein complex formed is of high affinity it may be stable to washing conditions which could desorb non-specifically bound proteins. Washing buffers should be intermediate between the best adsorption and elution conditions. Thus, if a protein binds in low molarity phosphate buffer and elutes in 0.75 M NaCl, it would be reasonable to try washing the column in 0.4 M NaCl to discover if any improvement can be made to the purity.

3.2.3 Choice of elution conditions

A protein molecule adsorbed to an affinity column is in equilibrium with the surrounding immobilized ligand (i.e. it is constantly undergoing desorption and adsorption events). If both the ligand concentration and the affinity are high it will be 'captured' by a ligand molecule at a short distance from its original position and show little, if any, movement through the adsorbent bed. Elution is effected by changing the environment of the ligand–protein complex such that the affinity of the ligand for the protein is lowered.

Most ligand–protein interactions are composed of a combination of molecular force (e.g. ionic, hydrophobic) and any solvent condition which sufficiently alters these will destabilize the ligand–protein complex. The destabilization should not, however, irreversibly denature the protein (if active protein is required) or the ligand (if column reuse is required). Often the nature of the binding interaction is not defined and elution conditions are found empirically. Continuous gradient elutions for up to 20 column volumes should be employed and compared for recovery and purity.

(a) **Change of ionic strength**. An increase in ionic strength (e.g. continuous gradient) is used to desorb proteins from ligands where ionic interactions predominate. Conversely, a lowering of ionic strength will be required to effect protein elution from adsorbent where hydrophobic interactions predominate. This is the most popular method of elution since it is inexpensive.

(b) **Change of pH**. This (generally downward) alters the degree of ionization of charged group at the binding surface so that it can no longer form a salt bridge with the opposing ions, thus, reducing the strength of interaction. In principle, an increase in pH should be, similarly, effective but is not as common in practice. This method is also inexpensive but gradients of pH are less reproducible and stepwise elution is more convenient.

(c) **Selective elution or affinity elution**. This uses molecules which are able to interact either at the ligand-binding site or at a different site, such that the binding surface is no longer available for binding (due to conformational change or steric occlusion). This is the mechanism by which affinity elution can be used for 'selective' desorption from ion exchange matrices. The biospecificity of the eluent does not necessarily imply biospecificity in the binding interaction.

A characteristic of affinity elution is that very low concentrations of eluent are required often less than 10 mM. Examples include elution of dehydrogenases from dye column by various nucleotides (1), or elution of glycoproteins from lectins by free sugars. Choice of selective eluents will be dictated by the availability and cost for each individual ligand–protein pair.

It should be noted that eluted protein may be difficult to separate from the eluent.

(d) **Chaotrophic agents**. These are used for elution when other methods fail, because of the very high affinity of an interaction. Chaotrophs effect desorption by disrupting the structure of water, thus, reducing ligand–protein interaction. Potassium thiocyanate (3 M), potassium iodide (2 M), or $MgCl_2$ (4 M) could be tried (see also Section 4).

(e) **Denaturants**. Urea (8 M) and guanidine–HCl (6 M) may also be effective. For both chaotrophs and denaturants their effect on protein activity should be examine before proceeding to scale-up.

(f) **Polarity reducing agents**. Dioxane, 10% (v/v) or ethylene glycol, 50% (v/v) should be tried. Similarly, low concentrations of detergents will also effect elution.

(g) **Other methods of facilitating elution**:
   (i)   Temperature gradient (22).
   (ii) Disruption of a ternary complex (23).
   (iii) Electrophoretic desorption (24).

(h) **Difficult elution or irreversible denaturation**. If either of these is experienced then reduction of the ligand concentration by dilution with underivatized matrix may permit the use of milder elution conditions. Alternatively, a lower concentration of immobilized ligand should be used.

3.3 Estimation of capacity

Having determined the conditions for effective adsorption and elution of the desired protein from an affinity adsorbent, it is necessary to determine adsorbent capacity before conditions for larger scale purification can be finalized. The best method, using a technique known as frontal analysis (25) is as follows:

(a) Prepare a 1 ml column (6 mm i.d.) of the selected adsorbent (unused) and equilibrate with the chosen buffer at 6–8 ml/h.

(b) Apply a constant stream of the crude protein mixture to be purified. Concentration of the desired protein P in the mixture is P_0. Collect 1 ml fractions.

(c) Monitor effluent for total protein (A_{280}) and assay fractions for the emergence of P.

(d) When the concentration of P in the effluent reaches P_0 adsorbent capacity is saturated. Wash the column until A_{280} reaches the baseline.

(e) Elute with chosen eluent. Measure the yield of P.

(f) Calculations (*Figure 1*): the volume at which the concentration of P in the effluent reaches 50% of P_0 is defined as V_e. Similarly, V_0 is defined as the volume at which the unadsorbed protein concentration is 50% of the original (this assumes that P is an insignificant proportion of the total protein).

The amount of P adsorbed to the column is $P_0 (V_e - V_0)$. This is the expected capacity. Ideally this should be the same as the 'working capacity' which is equal to the yield of P in the eluted fractions.

Note the following:

(a) If the working capacity is much lower than the expected capacity, elution conditions need to be improved.

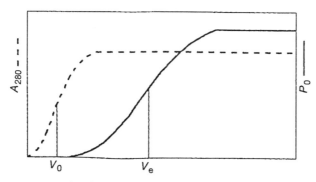

Figure 1 Estimation of the capacity of an affinity adsorbent. Breakthrough curve for total protein (broken line). Breakthrough curve for protein being purified (solid line). Figure prepared by D. Atwal.

(b) If assays for P cannot be accomplished in the time-frame of the experiment an 'educated guess' will have to be made as to the volume required. If the guess is wrong the experiment should be repeated, but on fresh unused adsorbent.

(c) If the flow rate is too high some P will appear in the effluent throughout the experiment.

(d) Capacity experiments should be carried out using conditions similar to those to be used in the actual purifications. Changes in the constituents of the feedstream will invalidate the capacity experiment, particularly, if the adsorbent binds several proteins whose proportions change.

(e) The capacity experiments should be repeated on the once-used adsorbent to determine the reusability of the column. This is particularly important for expensive affinity columns which are to be used repeatedly at more than 80% of the working capacity.

3.4 Ligand efficiency

Ligand efficiency may be defined as:

$$\frac{\text{Working capacity}}{\text{Theoretical capacity}} \times 100$$

The theoretical capacity is equal to the immobilized ligand concentration multiplied by the number of binding sites per molecule of ligand. In practice, ligand efficiencies of 1% or less are common for small ligands (4) indicating immobilization of ligand molecules in positions or orientations which are inaccessible to the interacting protein. On the other hand, macromolecular ligands, such as monoclonal antibodies (mAbs), often show efficiencies in excess of 10%. This observation probably reflects the much lower density of immobilization achievable with larger molecules. Thus, for successful affinity chromatography the aim should be to maximize ligand efficiency, particularly for expensive ligands, by experimenting with a range of ligand concentrations at the coupling stage.

3.5 Application to protein purification

3.5.1 Operational considerations

Previous sections of this chapter have dealt with the factors to consider in the design and preparation of affinity adsorbents as well as detailing the types of preliminary experiments required to be done before adaptation of such adsorbents for protein purification. This section briefly discusses the operational factors to be considered (see ref. 26 for a review.)

(a) **Mode of operation**. Batch mode is suitable for higher affinity systems, particularly when the volume of the starting material is high and concentration is not facile because of sensitivity of the protein of interest or because of high contaminant concentrations. Care should be taken to avoid excessively long incubation times which can contribute to secondary interactions.

Column mode is preferable for lower affinity systems since it is less labour-intensive and more flexible.

(b) **Column dimensions**. The volume of affinity matrix required will depend on the empirically found capacity and the type of separation required. For expensive ligands it is desirable to use more than 80% of the binding capacity. If the choice of ligand and elution conditions has resulted in good selectivity, resolution of proteins will depend on the adsorbent itself and not to any great degree on column length. Thus short, fat columns are quite suitable (e.g. diameter:length = 1:1) for higher affinity systems.

(c) **Flow rate**. This will depend on the porosity of the matrix and the size of protein to be purified and should, generally, be as slow as practicable. If the flow rate is too high the protein of interest will appear in the breakthrough before the column is saturated and will tend to cause tailing of peaks during the elution phase. As a guide, for agarose-based adsorbents, linear flow rates of 10 ml/cm^2/h can be recommended. Faster flow rates can be used during the washing and re-equilibration steps.

(d) **Ligand leakage**. A serious limitation of affinity chromatography is leakage of ligand into the feedstream or eluate (27). With high affinity systems (e.g. hormone–receptor interactions) the isolation of minute amounts of protein is compromised by the release of ligand from the matrix.

In the case of therapeutic proteins the presence of ligand in the final product is likely to result in serious consequences (29). Leakage may be due to instability of the immobilized ligand, instability of the linkage to the matrix, or dissolution of the matrix itself. All these factors will need to be investigated and their effects minimized in the development of therapeutic proteins.

Leakage of ligands in any substantial amount will lower the capacity of an adsorbent, which is unacceptable if the adsorbent is expensive and re-usability is essential for economy.

Ligands coupled to CNBr-activated matrices have been reported to show a greater degree of leakage than those coupled using other methods. However, since the instability of linkage is not the sole cause of leakage further work must be done on individual matrix–ligand systems where any degree of leakage is unacceptable.

(e) **Cleaning and storage of affinity adsorbents**. The robustness of an affinity adsorbent will depend on the nature of the ligand. For many systems, particularly those involving protein ligands, cleaning with extremes of pH or heat-sterilization is not possible. Affinity adsorbents are, therefore, often cleaned with high salt concentrations (2 M KCl) and stored in the presence of bacteriostatic agents (e.g. 0.02% (w/v) sodium azide).

3.5.2 Selected examples of use of affinity adsorbents

Some of the most commonly used ligands in affinity chromatography are lectins and antibodies. A brief summary of the use of other commonly employed ligands is given below. It should be used as a quick reference guide. More detailed information may be sought from refs 1–11 and manufacturer's publications.

(a) **Protein A** (29). Binds specifically to the Fc region of immunoglobulins from various species. Binds only weakly to murine IgG1, horse IgGc, chicken IgG, most IgA and IgMs. Does not bind human IgG3, or rat IgG2a and 2b. Binding is enhanced in the presence of high salt concentration, for example 3 M NaCl, and at high pH (e.g. 8–9). Usual binding buffer is phosphate or Tris based. Elution should be carried out with a decreasing pH gradient using 0.1 M citric or 1 M acetic acids. The ligand itself is stable to 6 M guanidine–HCl, which should be used to clean the column from time to time. Store in 70% ethanol.

(b) **Protein G** (30). Complements protein A in that it can bind human IgG3, rat IgG2a and 2b, as well as other IgGs mentioned above (with the exception of chicken IgG). Binding should be carried out in physiological buffers. Elution requires 0.1 M glycine–HCl pH 2.5. Eluted proteins should be neutralized immediately. Clean the column with 6 M guanidine–HCl at low pH and store in 70% ethanol.

(c) **Cibacron blue 3 G-A** (31). Binds human serum albumin, fibroblast interferon, lipoproteins, nucleotide-requiring enzymes (at least some are bound through their coenzyme-binding sites). Binding is usually in low molarity Tris–HCl pH 7–8.5 (but depends on the stability of the protein of interest). Binding is enhanced at lower pH, probably at least partially due to the ion exchange effect of the sulfonate groups on the dye. Elution is often successful using a salt gradient to 1 M KCl or alternatively affinity elution is possible. More stringent conditions are required for complete elution of human serum albumin (e.g. 0.5 M potassium thiocyanate). Columns can be cleaned with 8 M urea. Lipoproteins (e.g. from serum feeds) may clog the column surface and are best dealt with by unpacking the column into distilled water and decanting the lipoprotein particulate matter followed by several urea washes at extremes of pH. Leakage of ligand molecules is easily apparent due to their colour, particularly after storage, and adsorbents should be thoroughly washed to remove leached or non-covalently bound dye. A blank run with equilibration and elution buffers is recommended.

(d) **Other biomimetic ligands**. Since the discovery of the interaction of Cibacron blue with various enzymes there has been a considerable expansion in the application of other reactive dyes to protein purification. Dye–ligand or biomimetic chromatography has become popular because the ligands are inexpensive, easy to couple to matrices, and extremely stable. It is not usually possible to predict if a particular protein will bind to a given biomimetic column. It is, therefore, necessary to use systematic screening procedures (32, 33) to identify suitable adsorbents from a collection (e.g. kits available from Sigma and Prometic Bioscience). The general procedure for testing any dye adsorbents is that given for Cibacron blue above. Dye-ligand adsorbents can be very effectively used in tandem (33).

(e) **Phenyl boronate (PBA)**. PBA forms covalent complexes with *cis*-diols and may, therefore, be used to purify some glycoproteins (1). Binding is effected in 50 mM Tris–HCl pH 8 containing 20 mM $MgCl_2$. Elution requires sorbitol (0–50 mM gradient) in 50 mM Tris–HCl pH 8. A further wash at pH 5 using acetate buffer will clean and re-equilibrate the matrix.

(f) **Lysine**. Binds plasminogen, plasmin, and plasminogen activator (34). Binding is effected in physiological buffers and is followed by washing with 0.5 M NaCl in the same buffer. Specific elution may be carried out using 0.2 M ε-aminocaproic acid in distilled water.

(g) **Other applications**. The reader is referred to other texts in this series for additional practical information on applications (1, 35). Regular literature searches are prepared by Sturgen and Kennedy (36). Useful monographs and technical documents are also available from suppliers (e.g. Amersham Pharamacia, Biosepra, Perseptive).

4 Immunopurification

C. R. Hill
Lonza Biologics, 224 Bath Road, Slough, Berkshire SL1 4EN, UK.

L. G. Thompson
Consultant, The Cottage, The Common, Downley, High Wycombe, Buckinghamshire HP13 5YJ, UK.

A. C. Kenney
Oros Systems Ltd., Albion Close, Petersfield Avenue, Slough SL2 5DV, UK.

Immunopurification is one of the most selective and powerful purification methods available. Antibodies can be found that can distinguish between very similar antigens (37) and overcome separation difficulties that no other method can resolve. mAbs can be obtained by immunization with relatively impure antigen, thus, they can be obtained before alternative purification methods have been developed and so save time and effort in obtaining pure antigen. There may be problems associated with the removal of a particular contaminant from a relatively pure protein (for example, it is often difficult to remove the last traces of albumin from proteins purified from serum or serum-containing cell culture supernatants). In this case immunopurification can be used to subtract unwanted trace contaminants.

A study (38) showed that affinity chromatography is rarely used early in a purification procedure, where the high degree of purification often achieved can be exploited from both proteolytic attack and fouling by contaminants in the crude protein solution.

Immunopurification has been perceived as one of the most expensive affinity methods, particularly, when using mAbs. However, the expansion of mAb technology, including methods for large scale economic production (39, 40) has significantly reduced the costs of immunopurification, allowing the development of production scale applications (41–45).

In general antibodies as a class of proteins are, particularly, resistant to proteolytic attack and are cleaved quite selectively by only a limited number of enzymes. This can give immunopurification a distinct advantage over other affinity methods employing immobilized proteins and enables the method to be used early in a sequence of purification steps, where its power can best be exploited. Indeed immunopurification can often be considered the method of choice for purification of any protein for which a suitable mAb is available. Furthermore, mAbs are often raised against a new protein of interest early in a research programme to facilitate assay of the protein and identification by immunoblotting, thus antibodies are often available for use in purification.

Given the availability of a mAb it is usually a simple task to develop an immunopurification method that gives good results at least for research purposes. Many pre-activated support matrices are readily available (46, 47), so workers do not need to be exposed to the, often highly toxic, chemicals needed for the activation. These matrices can often be simply mixed with the antibody in a suitable buffer, then washed and blocked following a straightforward protocol. Some care is necessary to find a suitable eluent to remove the antigen of interest from the column; the stability of both the antigen and the antibody need to be borne in mind when selecting the eluent. Fortunately, most antibodies are remarkably robust to extremes of pH and ionic strength, and it is usually the nature of the antigen that dictates the composition of the eluent.

4.1 Antibody selection

The development of a successful immunopurification reagent is dependent upon the selection of a mAb with specific properties. Production of mAbs can often result in a large panel of antibodies. Several screening procedures are required to identify those antibodies with appropriate properties for development into an immunopurification reagent.

The steps in the development of an immunopurification reagent, at which decisions regarding the selection of an antibody can be made, are listed in *Table 4*.

Table 4 Criteria used for the selection of mAbs for development of an immunopurification reagent

Stage of development	Selection criteria	Typical number of cell lines in panel
1. Screening of hybridoma	Specificity, affinity	50–200
2. Hybridoma culture	Growth kinetics, nutritional requirements, cell line stability	10–15
3. mAb purification	Yield of antibody, stability of antibody	5–8
4. Immunopurification reagent development	Immobilization yield, elution buffer, capacity for antigen, yield of antigen, purity of antigen, reusability of reagent, leakage of IgG	4–6

Also given are a number of important criteria to be considered during the selection process.

4.1.1 Stage 1

A key step in the selection is identification of an antibody with an appropriate affinity and specificity for the antigen (48–50). A high affinity antibody should be chosen to allow efficient recovery of a dilute antigen from a complex mixture. However, with a high affinity antibody it may be difficult to recover the antigen. Thus, an antibody with an intermediate affinity may be more desirable (51).

It is not practicable to grow 50–200 cell lines in sufficient quantity to purify antibody from each. Screening assays that can be applied during the hybridoma development phase are, therefore, required, preferably using the supernatants from clones grown in 96-well microtitre plates. At its simplest this may involve determining their relative titres by titrating the antibodies against antigen immobilized on a microtitre plate.

The development of specific assays for screening hybridoma cell lines to identify antibodies with a suitable affinity for immunopurification has been described by Rubinstein and co-workers for immunopurification of α and β interferon (48, 59). The objective at this stage should be to identify perhaps 10–15 antibodies for further development.

4.1.2 Stage 2

The next stage in the selection procedure requires production of 10–100 mg of each antibody. This may be achieved by growth either as an ascites tumour or in suspension cell culture. If cell culture is used, information on the stability of the cell lines, their growth kinetics, and nutritional requirements can be generated. Those cell lines that exhibit low growth rates or poor antibody yields can be eliminated at this stage. mAbs from the remaining cell lines can then be purified using a variety of techniques (35).

4.1.3 Stage 3

At this stage information relating to yield and purity of antibody together with its isoelectric point, solubility, and stability can be obtained. Those mAbs that are difficult to purify or are unstable (52) can be eliminated at this stage.

4.1.4 Stage 4

Further selection will be based on properties after immobilization on a suitable matrix. Important criteria to be considered at this stage include yield of immobilized antibody, elution conditions required to recover antigen, capacity for antigen, yield and purity of antigen recovered, reusability of the reagent, and leakage of immobilized antibody.

4.2 Immobilization of antibodies on CNBr-activated Sepharose

For most research purposes, CNBr-activated Sepharose (Amersham Pharmacia) will be adequate. The material is readily available and convenient to use.

(a) Preparation of the antibody:
  (i) Dialyse the antibody at 4°C against fresh coupling buffer, either 0.1 M sodium hydrogen carbonate (NaHCO₃) pH 8.3 containing 0.5 M sodium chloride, or 0.1 M sodium borate buffer pH 8. As a guide, the amount of antibody required for immobilization is approximately 1–10 mg/ml gel for mAbs and about 10–20 mg/ml gel for polyclonal antibodies. These quantities vary according to the avidity of the antibodies for their antigens. Excess antibody may result in reduced efficiency due to steric hindrance of the immobilized antibody.
  (ii) Adjust the concentration of antibody in coupling buffer to 5–10 mg/ml. This can be estimated by measuring A_{280} using an average extinction coefficient for antibodies of 1.4 for 1 mg/ml protein using a 1 cm path length cell.

(b) Preparation of CNBr-activated Sepharose-4B for coupling antibody:
  (i) Chill a small volume (~ 10 gel volumes) of the chosen coupling buffer to 4°C and prepare a solution of 1 mM HCl (~ 200 ml/g gel).
  (ii) Weigh out the desired amount of CNBr-activated Sepharose-4B powder immediately prior to use (0.3 g produces ~ 1 ml of swollen gel). The powder should not be allowed to stand for any length of time in the open laboratory as it absorbs moisture from the atmosphere, which can destroy the CNBr-activated sites.
  (iii) Sprinkle the dry powder slowly onto 10 volumes of 1 mM HCl contained in a sintered glass funnel.
  (iv) Stir the gel gently with a spatula to ensure that all the particles are dispersed. The gel particles are extremely fragile and should be treated gently. Shear forces caused by magnetic stirrer bars and vigorous manual stirring can cause the particles to fracture, producing fines which adversely affect ultimate flow rates, and can block sinters and column support nets.
  (v) Allow the gel to swell at room temperature for approximately 15 min and then flush through with the remaining 1 mM HCl under suction, taking great care not to allow the gel to dry, as this can also break down the matrix. A total of 200 ml of 1 mM HCl per gram of dry gel is sufficient for this washing step.
  (vi) Finally wash the gel with 10 gel volumes of chilled (4°C) coupling buffer.

(c) The coupling reaction:
  (i) Mix one volume of washed gel with five volumes of antibody solution in coupling buffer at 4°C. It is important that both reactants are at 4°C prior to mixing, as at higher temperatures the CNBr-activated groups react faster with the amino groups on proteins. This can cause dense coupling of antibody molecules to the external surfaces of the gel and result in reduced efficiency.
  (ii) Incubate the coupling mixture at 4°C for 16–20 h; continual mixing is necessary to prevent the gel settling and a rolling motion rather than an end-over-end motion is more gentle on the gel beads.

(d) Blocking and washing the Sepharose gel. When the coupling reaction is complete it is advisable to test the success of the reaction before proceeding.

(i) Allow the mixture to settle for a few minutes.

(ii) Take a sample (~ 1 ml) of the supernatant, centrifuge to clarify and measure the A_{280} of the supernatant; an absorbance close to zero suggests a successful coupling reaction.

(iii) Add this supernatant back to the reaction mixture.

(iv) Filter the coupling reaction mixture by gentle suction in a sintered glass funnel, and wash the gel with a further five volumes of fresh coupling buffer.

(v) Collect all filtrates in one vessel and determine protein content. Do not allow the gel to dry.

Any remaining CNBr-activated groups on the matrix are then blocked by adding an excess of a primary amino-containing compound. A variety of compounds may be used for this purpose (e.g. ethanolamine, glycine) and each should be considered in the light of the property it may impart to the final reagent. For example, ion exchange properties or hydrophobic character could enhance the non-specific binding of contaminant proteins to the reagent. The most commonly used blocking agent is 1 M ethanolamine adjusted to pH 8 with HCl. Care must be taken in the preparation of this solution, since acidification requires a high volume ratio of concentrated hydrochloric acid to be added to the ethanolamine, and the reaction is exothermic. The following procedure should be followed:

(a) Incubate the antibody matrix with blocking solution at room temperature for 1–2 h with continual rolling.

(b) Remove excess ethanolamine from the gel using a sintered glass funnel and gentle suction, taking care not to dry the gel out.

(c) Wash the gel in three alternating cycles of high and low pH buffers (0.1 M sodium acetate buffer pH 4 containing 0.5 M sodium chloride; and 0.1 M sodium carbonate buffer pH 8.3 containing 0.5 M sodium chloride). These washing cycles promote the release of non-covalently bound antibodies. Use a sintered funnel for the washing and allow the gel to stand at the appropriate pH for 5–10 min during each cycle. Use ten gel volumes of each wash solution.

(d) A further series of washing cycles to minimize eventual antibody leakage can be introduced at this point. For this, the equilibration buffer and elution buffer (see below) are used.

(e) Store the gel at 4°C in equilibration buffer with 0.1% (w/v) sodium azide added. Once swollen, the gel should never be frozen.

A convenient means of aliquoting known volumes of gel is as follows. Allow the final mixture to settle for approximately 72 h before adjusting the volume of liquid on top of the gel to equal that of the gel. Each time gel is required from the container it can be gently mixed to form an even slurry and two times the desired volume of settled gel removed.

4.3 Procedures for immunopurification

There are two ways of employing immunopurification. First, in a positive fashion, to purify the antigen from a crude mixture, and secondly, in a negative (subtractive) fashion, to remove a specific contaminant from a partially purified antigen solution. There are also two ways of executing the method, that is either by a batch method or a column procedure.

4.3.1 Positive immunopurification

In this method the immobilized antibody is specific for the protein to be puri-fied. It is, therefore, important to determine under what conditions the antigen binds to the immunoadsorbent and what conditions are necessary to effect dissociation of the complex. All conditions applied in this step should be com-patible with both the immunoadsorbent and the antigen.

i. Equilibration

The equilibration buffer is one in which the antigen binds to the immobilized antibody. It is usually a neutral, low ionic strength buffer such as 40 mM sodium phosphate buffer pH 7.2, or phosphate-buffered saline. The immunoadsorbent is stored in this buffer containing 0.1% sodium azide, and equilibrated in fresh buffer, without azide, when required for use. The immunoadsorbent should be equilibrated to the same temperature as the equilibration buffer and the environment in which the column is to be used; otherwise air bubbles can form and be trapped if the matrix is already packed in a column. Pack the immuno-adsorbent in an appropriate column. Equilibrate in ten or more volumes of equilibration buffer (to remove traces of azide used in storage and to flush out any ligand, which may have leached on storage). The column is now ready for sample application.

ii. Sample loading

Ensure the crude antigen sample is in equilibration buffer, for example by dialysis, and allow it to equilibrate to the same temperature as the packed column. If the antigen solution contains any particulate matter (i.e. protein precipitate or cell debris) clarify by centrifugation or filtration immediately prior to loading or pass via a 0.22 μm filter onto the column. This procedure protects the immuno-adsorbent and prevents column blockage. The flow rate should be as slow as convenient and no faster than that recommended for the Sepharose-4B matrix (~ 30 ml/cm^2/h). A slow flow rate allows all antigen molecules passing through the column to saturate all available binding sites, whereas a faster flow rate risks losing some antigen in the flow-through before capacity is reached.

The capacity of an immunoadsorbent is defined as the maximum amount of antigen that can be bound and recovered per unit volume and should be experimentally determined for each immunoadsorbent.

Theoretically each monoclonal IgG antibody immobilized can bind two antigen molecules. However, in practice only a small percentage of antigen binding sites remain available for binding antigen after covalent immobilization to a matrix (53). Typically this figure will be, approximately, 10% of available antigen binding sites. However, this can vary over a wide range (< 1–50). A number of factors will affect the apparent efficiency of the matrix. These include coupling chemistry, matrix pore size, antigen size, and antibody structure. For example, an antibody that contains a reactive amino group in the antigen combining site may well result in a matrix with a very low overall efficiency.

iii. Washing

Once the antigen has bound to the immunoadsorbent it is necessary to wash off the non-specifically bound contaminant molecules prior to elution. This may be done by passing equilibration buffer through the column until the absorbance at 280 nm of the effluent is about 0 (~ 10 gel volumes). If the Ab–Ag complex is sufficiently strong, inclusion of, approximately, 0.5 M sodium or potassium chloride in the equilibration buffer may result in a purer product.

iv. Elution

Elution of bound antigen from the immunoadsorbent is effected by breaking the bonds which from the complex. These bonds will be composed of a mixture of weak physical forces such as coulombic salt bridges, hydrogen bonds, and Van der Waals forces.

Elution is usually achieved by a change in ionic strength, pH, dielectric constant, surface tension, or temperature. However, to preserve the antigen and immunoadsorbent activity the mildest conditions should be tried before the more stringent conditions. Thus, the order in which to try eluents can be summarized as increased ionic strength followed by extremes of pH, chaotropic agents, protein deforming agents, and organic solvents (*Table 5*). Since an elution

Table 5 Some elution conditions for immunopurification in the order in which they should be tried

Elution conditions	References
Extremes of pH	
0.1 M glycine–HCl pH 2.5	54
1.0 M propionic acid	55
0.15 M NH$_4$OH pH 10.5	56
0.1 M NaCaps pH 10.7	42
Chaotropic salts	
4.0 M MgCl$_2$ pH 7.0	57
2.5 M NaI pH 7.5	58
3.0 M NaSCN pH 7.4	59
Organic chaotropes	
8 M urea pH 7.0	60
6 M guanidine–HCl	61
Organic solvents	
50% (v/v) ethanediol pH 11.5	62, 63
Dioxane/acetic acid	62, 63

step requires only two gel volumes of elution buffer, elution can be achieved in a relatively short time; thus, if the elution condition is particularly harsh a rapid change over to the re-equilibration step and/or a rapid buffer change for the purified antigen can reduce damage caused by elution. This can be achieved either by collecting the eluent into tubes containing a neutralizing buffer or by rapid gel filtration.

4.3.2 Example of positive immunopurification of interleukin 2 (IL-2)

A mAb to IL-2 was immobilized on CNBr-activated Sepharose at a concentration of 8 mg IgG/ml packed gel. The immunoadsorbent was packed into a suitable column and equilibrated with 0.1 M sodium phosphate buffer pH 7.5 containing 0.5 M sodium chloride. Crude IL-2 derived from *E. coli* lysate was loaded onto the column at 100 ml/cm^2/h. The column was washed with ten gel volumes of 0.1 M sodium phosphate buffer pH 7.5 containing 1.0 M potassium chloride. IL-2 was eluted from the immunoadsorbent with five gel volumes of 0.1 M sodium acetate buffer pH 4.0 containing 1.0 M potassium chloride. The eluent was neutralized by collecting directly into 2.5 gel volumes of 2 M Tris–HCl pH 7.5 and the immunoadsorbent was immediately re-equilibrated in equilibration buffer containing 0.1% azide.

Electrophorectic analysis of various fractions can be seen in *Figure 2*. A purification of about 1000-fold was achieved in this single step.

4.3.3 Negative (subtractive) immunopurification

In this method the immobilized antibody is specific for a particular contaminant. The principles described in the previous section for positive immunopurification still apply but the emphasis is on different parts of the method.

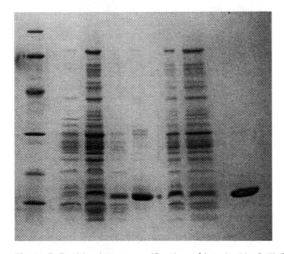

Figure 2 Positive immunopurification of interleukin 2 (IL-2). Coomassie Blue-stained 7.5–15% SDS–PAGE run under reducing conditions. Tracks from left to right: mol weight standards, sample load flow-through (and five volumes), column wash (one and five volumes), eluted IL-2 (one and five volumes), crude IL-2 *E. coli* lysate (one and five volumes), IL-2 standard. Photograph provided by D. Brady.

(a) **Equilibration**. The main criteria for selection of an equilibration buffer are suitability to allow or promote the antigen binding to the immunoadsorbent and compatibility with the desired protein. This buffer will usually be a neutral, low ionic strength buffer. The column should be packed and equilibrated in this buffer as described previously.

(b) **Purification**. Prior to application to the immunoadsorbent column, dialyse the extract into equilibration buffer. To be sure of removing all of the contaminant it is important not to exceed the capacity of the immunoadsorbent and to use an appropriate flow rate to allow contaminant antigens to bind.

(c) **Column regeneration**. The immunoadsorbent can be recycled many times if it is regenerated carefully after each use. This step can be compared with the elution step for positive immunopurification, since the same principles apply. However, unless the contaminant antigen is required, it is less important to consider its compatibility with the elution buffer. Apply two to five gel volumes of elution buffer to the immunoadsorbent followed by ten or more gel volumes of equilibration buffer. Store the immunoadsorbent in equilibration buffer containing 0.1% sodium azide.

4.3.4 Example of negative immunopurification

An mAb to BSA was immobilized onto CNBr-Sepharose at 5 mg IgG/ml packed gel. The immunoadsorbent was packed into a suitable column with phosphate-buffered saline (PBS). Two gel volumes of elution buffer (0.1 M glycine–HCl

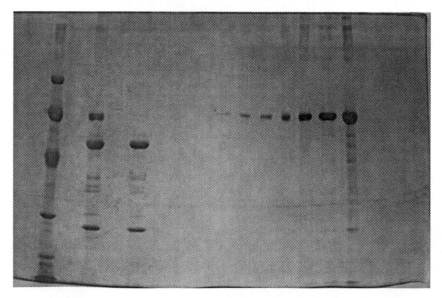

Figure 3 Negative immunopurification of bovine serum albumin (BSA) contamination from an mAb. Silver stained 7.5–15% SDS–PAGE run under reducing conditions. Tracks from left to right: mol.wt. standards, sample before purification, sample after purification, BSA standards: 0, 5, 10, 25, 50, 100, 500 ng, 1 μg. Photograph provided by A. Nash.

pH 2.5) were passed through the column as a pre-wash. The column was then equilibrated with ten or more gel volumes of PBS.

A sample of an mAb containing BSA was applied to the immunoadsorbent column at 10 ml/cm²/h. The unbound proteins were collected and analysed by SDS electrophoresis, together with the sample applied (*Figure 3*). An estimate from the electrophoresis gel of BSA clearance shows a reduction in BSA from 10% (w/w) to less than 0.05%, demonstrating at least a 200-fold reduction in BSA in a single pass.

References

1. Dean, P. D. G., Johnson, W. S., and Middle, F. A. (ed.) (1985). *Affinity chromatography: a practical approach*. IRL Press, Oxford.
2. Jakoby, W. (ed.) (1984). *Methods in enzymology*, Vol. 104C. Academic Press, New York.
3. Scouten, W. H. (1981). *Affinity chromatography*. Wiley-Interscience, New York.
4. Lowe, C. R. (1979). *An introduction to affinity chromatography*. Elsevier Biomedical, Amsterdam.
5. Lowe, C. R. and Dean, P. D. G. (1974). *Affinity chromatography*. John Wiley & Sons, London.
6. Jackoby, W. (ed.) (1974). *Methods in enzymology*, Vol. 34. Academic Press, New York.
7. Turkova, J. (1978). *Affinity chromatography in J. Chromatography Library*, Vol. 12. Elsevier, Amsterdam.
8. Jennissen, H. P. and Muller, W. (ed.) (1988). *Die Makromol. Chem. Macromol. Symp.*, **17**.
9. Chaiken, I. M., Wilchek, M., and Parikh, I. (ed.) (1983). *Affinity chromatography and biological recognition*. Academic Press, New York.
10. Gribnau, T. C., Visser, J., and Nivard, R. T. F. (ed.) (1982). *Affinity chromatography and related techniques*. Elsevier, Amsterdam.
11. Hoffman-Ostenoff, O., Breitenbach, M., Koller, F., Kraft, D., and Scheirier, O. (ed.) (1978). *Affinity chromatography*. Pergamon Press, New York.
12. Fowell, S. J. and Chase, H. A. (1986). *J. Biotechnol.*, **4**, 355.
13. Angal, S. and Dean, P. D. G. (1977). *Biochem. J.*, **167**, 301.
14. Kohn, J. and Wilchek, M. (1982). *Biochem. Biophys. Res. Commun.*, **107**, 878.
15. Bethell, G. S., Ayers, J. S., Hancock, W. S., and Hearn, M. T. W. (1979). *J. Biol. Chem.*, **254**, 2572.
16. Nilsson, K. and Mosbach, K. (1980). *Eur. J. Biochem.*, **112**, 397.
17. Ngo, T. T. (1986). *Bio/Technology*, **4**, 134.
18. Weston, R. D. and Avrameas, S. (1971). *Biochem. Biophys. Res. Commun.*, **45**, 1574.
19. Avrameas, S., Ternynck, T., and Guesdon, J. (1978). *Scand. J. Immunol.*, **8**, 7.
20. Lowe, C. R., Harvey, M. J., and Dean, P. D. G. (1974). *Eur. J. Biochem.*, **41**, 341.
21. Scopes, R. K. (1982). In *Affinity chromatography and related techniques. Analytical Chemistry Symposium Series* Vol. 9, p. 333.
22. Harvey, M. J., Lowe, C. R., and Dean, P. D. G. (1974). *Eur. J. Biochem.*, **41**, 353.
23. Hughes, P., Lowe, C. R., and Sherwood, R. F. (1982). *Biochim. Biophys. Acta*, **700**, 90.
24. Morgan, M. R. A., Slater, N. A., and Dean, P. D. G. (1978). *Anal. Biochem.*, **92**, 144.
25. Nicol, L. W., Ogston, A. G., Winzor, D. J., and Sawyer, W. H. (1974). *Biochem. J.*, **143**, 435.
26. Osterman, L. A. (1986). *Methods of protein and nucleic acid research. Part 3. Chromatography*, p. 308. Springer–Verlag, New York.
27. Tesser, G. I., Fisch, H.-U., and Schwyzer, R. (1974). *Helv. Chim. Acta*, **57**, 1718.
28. Bristow, A. (1989). In *Protein purification applications: a practical approach* (ed. E. L. V. Harris and S. Angal). IRL Press, Oxford.

29. Kruger, N. J. and Hammond, J. B. W. (1988). In *Methods in molecular biology* (ed. J. M. Walker), Vol. III, p. 363.

30. Frederiksson, G., Nilsson, S., Olsson, H., Bjorck, L., Akerstrom, B., and Belfrage, P. (1987). *J. Immunol. Methods*, **97**, 65.

31. Dean, P. D. G. and Watson, D. H. (1979). *J. Chromatogr.*, **165**, 301.

32. Scopes, R. K. (1986). *J. Chromatogr.*, **376**, 131.

33. Hey, Y. and Dean, P. D. G. (1983). *Biochem. J.*, **109**, 363.

34. Radcliffe, R. and Heinze, T. (1978). *Arch. Biochem. Biophys.*, **189**, 185.

35. Harris, E. L. V. and Angal, S. (ed.) (1989). *Protein purification applications: a practical approach*. IRL Press, Oxford.

36. Sturgeon, C. M. and Kennedy, J. F. *Enzyme and microbial technology* (regular feature).

37. Jack, G. W., Blazek, R., James, K., Boyd, J. E., and Micklem, L. R. (1987). *J. Chem. Technol.*, **39**, 45.

38. Bonnerjea, J., Oh, S., Hoare, M., and Dunnhill, P. (1986). *Bio/Technology*, **4**, 954.

39. Birch, J. R., Thompson, P. W., Lambert, K., and Boraston, R. (1985). In *Large scale mammalian cell culture* (ed. J. Feder and W. R. Tolbert). Academic Press, London, New York.

40. Birch, J. R., Lambert, K., Thompson, P. W., Kenney, A. C., and Wood, L. A. (1987). In *Large scale cell culture technology* (ed. B. J. Lyderson), p. 1. Carl Hanser Verlag, Munich.

41. Janson, J.-C. (1986). *Trends Biotechnol.*, **2**, 31.

42. Hill, C. R., Birch, J. R., and Benton, C. (1986). In *Bioactive microbial products* (ed. J. D. Stowell, P. J. Bailey, and D. J. Winstanley), Vol. 3, p. 175. Academic Press. London and New York.

43. Secher, D. S. and Burke, D. C. (1980). *Nature*, **235**, 446.

44. Muller, H. P., Van Tilbrug, N. H., Derks, J., Klein-Breteler, E., and Bertine, R. M. (1981). *Blood*, **58**, 1000.

45. Tarnowski, J. J. and Liptak, R. A. (1983). In *Advances in biotechnological processes*, Vol. 2, p. 271. Alan R. Liss Inc., New York.

46. Kenney, A. C., Lee, L. G., and Hill, C. R. (1988). In *Methods in molecular biology* (ed. J. M. Walker), Vol. 3, p. 99. Humana Press Inc.

47. Hill, C. R., Kenney, A. C., and Goulding, L. (1987). *BIF-Biotech Forum*, **4**, 167.

48. Novick, D., Eshhar, Z., and Rubinstein, M. (1982). *J. Immunol.*, **129**, 2244.

49. Novick, D., Eshhar, Z., Gigi, O., Marks, Z., Revel, M., and Rubinstein, M. (1983). *J. Gen. Virol.*, **64**, 905.

50. Cobbs, C. S., Graus, P. K., Russ, ?, *et al.* (1983). *Toxican*, **21**, 285.

51. Chase, H. A. (1981). *Chem. Eng. Sci.*, **39**, 1099.

52. Underwood, P. A. and Bean, P. A. (1985). *J. Immunol. Methods*, **80**, 189.

53. Chase, H. A. (1984). *J. Biotechnol.*, **1**, 67.

54. Hudson, L. and Hay, F. C. (1980). *Practical immunology*. Blackwell Scientific, Oxford.

55. Kristianson, T. (1978). In *Affinity chromatography* (ed. O. Hoffman-Ostenhoff, M. Breitenbach, F. Koller, D. Kraft, and O. Scheiner). Pergamon, New York.

56. Chidlow, J. W., Borune, A. J., and Bailey, A. J. (1974). *FEBS Lett.*, **41**, 248.

57. Mains, R. E. and Eipper, B. A. (1976). *J. Biol. Chem.*, **251**, 4115.

58. Avrameas, S. and Ternynck, T. (1967). *Biochem. J.*, **102**, 37C.

59. Zoller, M. and Matzku, S. (1976). *J. Immunol. Methods*, **11**, 287.

60. Melchers, F. and Messer, W. (1970). *Eur. J. Biochem.*, **17**, 267.

61. Weintraub, B. D. (1970). *Biochem. Biophys. Res. Commun.*, **39**, 83.

62. Hill, R. J. (1972). *J. Immunol. Methods*, **1**, 231.

63. Anderson, K. K., Bejamin, Y., Douzov, P., and Bolny, C. (1979). *J. Immunol. Methods*, **25**, 375.

Chapter 10
Scale-up considerations

John B. Noble

Foster Wheeler Ltd., Foster Wheeler House, Station Road, Reading, Berkshire RG1 1LX, UK.

1 Scale-up of purification processes and unit operations

Once a potential indication has been found for a given protein product, it is necessary to produce increasing quantities to satisfy the demands of market trial activities. In the majority of cases, this will lead to scale-up of the purification process to meet demand whilst maintaining the safety, efficacy, and quality of the product. The following chapter provides an overview of the key issues that will arise during scale-up and provides the reader with practical advice on process and equipment selection.

The text examines the key issues in defining production scale, identifies critical scale-up and development issues on an overall process basis, presents practical tips on scale-up, and concludes with two industrial scale-up case studies.

1.1 Choice of scale

The final choice of purification scale must reflect the most cost-effective solution for the whole of an organization and as such there are a great many influences to be considered. As a result the development process is an iterative exercise in which the production demand and schedule are balanced against available resources. A step-by-step approach to choice of purification scale is developed below:

(a) **Step 1**: define volume of product required and when it is needed.

(b) **Step 2**: develop a preliminary scale-up schedule.

(c) **Step 3**: match scale and production schedule to production resources.
   Each of these steps will now be developed further.

1.1.1 Step 1: define volume of product required and when it is needed

From preliminary product trials it will be possible to develop a schedule of product demand against time which can be used as the building block for step 2. Typically for pharmaceutical products the schedule will included materials for pre-clinical trials, phases I–III clinical trials, and commercial manufacture. During trials a defined quantity of product will be required over a clearly defined period whilst, once a product has been approved, demand will be less well defined and generally increase gradually over time to market saturation.

1.1.2 Step 2: develop a preliminary scale-up schedule

From the schedule developed in step 1 and a knowledge of the approximate process yields, a preliminary assessment of overall raw material throughput can be developed. This can then be broken down and combined with information on product shelf-life to assess the most appropriate production strategy for each stage of a product's life. The following points are key to the assessment process:

(a) Minimize the number of scale-up steps as each scale-up will necessitate a degree of revalidation.

(b) Minimize the size of each scale-up operation to minimize the risk to production.

(c) Delay work requiring significant investment until as late as possible in a product's life to minimize financial risk.

By balancing scale and throughput (i.e. number of batches) a preliminary scale-up philosophy can be developed. This philosophy must then be considered as part of overall production strategies as outlined in step 3.

1.1.3 Step 3: match scale and production schedule to production resources

The philosophy developed in step 2 is product-specific and takes no account of available production resources. Most biotechnology companies have a wide portfolio of products in development, all of which compete for limited resources. Any scale-up strategy must thus take due account of these issues.

In the first instance, it must be assessed whether the preliminary scale-up schedule fits with existing resources. Most importantly the compatibility of scale, capacity, and equipment items/facility suitability must be checked. If the initial match is poor, then ways of revising the scale-up schedule should be investigated to see if scale or throughput can be tailored to fit resources. Although such an approach may not be cost-effective for a particular product it could prove more viable for production as a whole when compared with the significant investment required for new facilities and equipment. When assessing resources attention should be paid to in-house and external contract production.

Where large volumes of product (i.e. hundreds of kilograms per year) are required for trials great care must be taken in assessing scale-up options as dedicated production facilities are likely to be required.

1.2 Overview of critical issues

By working through the various steps outlined in Section 1.1, a clear picture of the required scale of operation can be developed. Once this is in place a series of options can be developed of how to reach this scale of manufacture involving different unit operations and process sequences. Each of these options must be challenged against a series of criteria to highlight the most promising scenario. The key scale-up criteria are as follows:

- safety
- robustness and reliability
- scale-up and scale-down
- product consistency
- future demands
- schedule

Taking each of the above categories, the process options must be challenged to arrive at the most cost-effective solution which delivers a safe product with the required efficacy. The scale-up team must be continually aware of the financial implications of its decisions. Of prime importance throughout is the concept of Keep it simple and an appropriate level of sophistication must be chosen which balances the time available for scale-up and the demands of the separation process.

The following text discusses each of the guide words highlighted above in more detail.

1.2.1 Safety

The safety of a manufacturing process is of prime importance. The more hazardous a production process, the more expensive it will be to implement at manufacturing scale operation. Areas for concern within biopharmaceutical production are as follows:

(a) **Biocontainment**. If recombinant or pathogenic organisms are involved in the fermentation then downstream operations must be reviewed to ensure that they do not compromise containment.

(b) **Pressure**. The use of high pressure processes should be avoided if possible and, if required, shifted to the latter stages of a process, where volumes will be smaller and safety issues more manageable.

(c) **Flammability**. The design of equipment and facilities for hazardous area operation has significant financial implications and should be avoided if possible. If required, this too should be shifted to the latter stages of a process, where volumes will be smaller and safety issues more manageable.

1.2.2 Robustness and reliability

Any process step used within the process must be both robust and reliable and hence validatable (1). In practical terms this means it must produce a repeatable quality of product over extended operation. If this is not the case then the quality, and hence safety, of the product cannot be guaranteed. Attention should focus upon the sensitivity of a given operation to process parameters such as pH, temperature, conductivity, and residence time and an assessment made of the practicalities of controlling these variables within the required ranges.

1.2.3 Scale-up and scale-down

The driving force behind scale-up of processes is the concept of scale-up and scale-down (2). Inherent within this approach is the need to use equipment items with similar operating principles but differing capacity at varying scales of operation. For example, if centrifugation is required and continuous operation is preferred, the scale-up team should try to source units which can be effective at all required scales. This enables clear definition of operating practices during early scale-up and minimizes work to ensure product consistency with scale-up. Conversely, if the design and operating principle of the unit operations has to change during scale-up, this will extend the project schedule and add cost. Equipment selection must thus take a detailed account of what will be required at later scales.

1.2.4 Product consistency

As the production process matures through various stages it is essential that the operator can demonstrate product consistency from scale to scale. This requires a significant amount of work at each increase in scale and it is thus prudent to minimize the number of times a process is scaled-up. Typically, a maximum scale-up factor of tenfold (2) is acceptable to the medicines inspectors worldwide but this must be assessed on a case-by-case basis. As discussed, the use of the scale-up scale-down philosophy can be of great help in demonstrating product consistency and forms a key aspect of compliance.

1.2.5 Future demands

The principle of scale-up and scale-down looks in detail at the process aspects of the unit operations. Outside of this there is the specification of the production facility itself and the selection of unit operation can have a dramatic effect upon this. A classical example is the consideration of building finishes and air conditioning. At small scale, it is easy to build specialist areas which afford the necessary protection to the production process. As scale increases this becomes more and more complex and costly. A key philosophy to adopt is the concept of closing the process which ensures that product protection is provided by the primary process envelope and eliminates the potential interaction between room environment and product stream. It is essential that facility requirements are not built in at small scale which will make the whole process uneconomic at production scale.

1.2.6 Schedule

Time is critical in getting products to market. Recent reports suggest that, for a therapeutic protein product grossing sales of $300–$400 million/year, $1 million can be lost for every day of delay to market launch. During scale-up, we are thus looking for an approach which ensures that scale-up activities do not lie on the critical path. This necessitates a pragmatic approach to process development and a heavy reliance upon tried and tested operations. The development of novel separation strategies can only take place if it is de-coupled from the critical path.

1.3 Practical aspects of scale-up

The guide words within Section 1.2 provide an overview of the key scale-up issues and allow the development of a generic process flow scheme. The next stage in the scale-up process is to go into detail with each area and consider specific scale-up topics. The following section pulls together the thoughts of various individuals within the biotechnology industry to provide a practical guide to the key points. The listings are not intended to be exhaustive but do provide the reader with prompts for the study of their own specific processes.

1.3.1 Homogenization

There are two main contenders for large scale cell disruption: bead mills and high pressure homogenization. The relative merits of the two steps are listed in *Table 1*.

In general, high pressure homogenization should be used unless breakage is unsuccessful. When scaling-up it is important to study product loss during disruption and ensure that operations are optimized to maximize recovery for the whole process. Tests have shown that laboratory scale homogenization experiments can be readily scaled-up (3).

1.3.2 Centrifugation

Centrifuges are used widely throughout biotechnology for solids removal operations. These operations can involve the clarification of a liquid product stream or the de-watering of a solid phase product. Typically, the unit will be used early in the purification process. Within the biotechnology industry, centrifuges are used for fairly onerous duties with small particle size and low density differences. Given these operating conditions, centrifugation often forms a significant proportion of the capital budget and can represent a bottle-neck in processing. It is essential that the optimum centrifugation train is selected. In the main, the choice is between fixed bowl and continuous discharge machines. The two are compared in *Table 2*.

Table 1 Comparison of cell disruption equipment

Attribute	Bead mill	High pressure homogenizer
Cleaning	Complex internals present cleaning problems	Relatively straightforward
Containment	Standard techniques apply	High pressure may require dedicated room environment
Ease of optimization	Many variables to study	Relatively straightforward
Contamination	Bead fragments may enter product stream	N/A

245

Table 2 Relative attributes of centrifugation systems

Attribute	Fixed bowl	Continuous discharge
Bowl speed	Up to 17 000 g	10–15 000 g
Closed/open processing	Processing must be open	Closed processing possible
Solids capacity	~ 10 litres	Continuous discharge
Throughput	Low unless multiple machines are used	High
Liquid solid separation	High	Medium
Solids shear	Low	Medium

The following points can be used as a guide for scale-up based upon *Table 2* and other operating experience:

(a) Where high g force processing is required, fixed bowl machines may be the only practical option unless the use of continuous centrifugation followed by filtration is viable.

(b) The need for open processing with fixed bowl machines may require specialist environments and this must be factored into cost models.

(c) The low solids capacity of the fixed bowl machines can necessitate multiple machines running in series at larger scales which has a significant impact on equipment and building costs.

(d) Solids ejection from continuous machines places a level of shear upon the process and the impact of this upon product integrity this must be assessed on a case-by-case basis.

(e) The complexity of centrifugal separators presents a series of unique cleaning challenges. In most cases, this will require a unique cleaning procedure based on a detailed assessment of process and facility. The integration of this procedure within the overall manufacturing strategy is critical.

(f) The complex nature of biological systems can lead to marked changes in sedimentation and rheological properties with only slight process changes. It is essential that performance is fully evaluated before commitment is made to a given process route. This must include real system equipment trials.

(g) The effect of product loss within liquid or solid phases must be assessed on a case-by-case basis. Fixed bowl machines give a dryer solid phase than continuous machines as the solid phase must be able to flow from the latter.

1.3.3 Chromatography

Chromatography is the mainstay of many modern biopurification processes and there is a great deal of experience in the design selection and scale-up of chromatographic separations. Axial flow columns are now well established within industry and, in recent years, radial flow systems (4) have begun to demonstrate increasing competitiveness. The relative attributes of radial and axial flow chromatography systems are given in *Table 3*.

Table 3 Relative attributes of chromatography systems

Attribute	Axial flow	Radial flow
Column size	Diameter up to 2 m Volume ~ 500 litres	Diameter up to 0.5 m Volume 350 litres
Throughput	High	Very high
Packing	Can be pump packed	Can be pump packed
Scalability	Direct scale-up based upon residencetime and linear velocity	Bed geometry can change with scale (100–350 litres) and performance may require revalidation
Operating experience	Extensive experience at all scales	Limited experience at largest scales

The following points can be used as a guide for scale-up based upon *Table 3* and other operating experience:

(a) As scale increases, radial flow systems become more ergonomic and easier to manage within facility design.

(b) For large scale high throughput production, radial flow systems may offer advantages but this must be assessed in terms of the whole process. If chromatography is not the bottle-neck the advantage may not be significant.

(c) Pressure drop in radial flow columns for equivalent throughputs are lower than axial systems.

(d) In certain radial flow systems the bed geometry changes as scale increases. This alters the velocity profile within the bed and thus the separation performance. Revalidation is thus required which will take up time and resource and must be planned.

(e) There is limited experience of operating radial columns at volumes at and above 350 litres whilst axial systems have been more widely used at such scales. Both systems have problems at these scales including bed stability and packing homogeneity.

(f) Adsorbents can represent a significant proportion of the process consumable costs and efforts should be made to maximize adsorbent lifetimes, 100 cycles are not uncommon within commercial manufacture. Care must be taken in assessing up stream operations to balance the impact these may have upon adsorbent life.

(g) To minimize adsorbent inventory, attention should focus upon multiple cycling of smaller columns rather than the purchase of a single large volume column.

One important point is that the choice of column geometry should be made at an early stage. It is always possible to change but to do so will require a complete revalidation of the separation performance. In the case of mammalian-derived products this could be further complicated by the need to validate viral removal.

The use of HPLC is becoming more prevalent in protein purification operations. During scale-up, attention must be paid to the need for flame-proof design and drainage dedicated to solvent handling.

1.3.4 Filtration

Once the flux across a given filter/membrane area has been established the scale-up of these operations is relatively straightforward. Membrane area is linearly increased to accommodate increased process flow. As scale increases, the proportion of disposable cost associated with filtration operations increases markedly and it is essential filter use is optimized. The following points should be considered during scale-up:

(a) Solids removal duties may be best handled with centrifugation followed by clean up via depth filtration.

(b) Multiple use membrane filters should be cleaned and stored such that their lifetime is maximized.

(c) Single use membrane filters, such as those for viral removal, should be used to full capacity. The pooling of batches should be considered to maximize filter throughput.

(d) The loss of product on filter medium whether by adsorption or degradation must be assessed in detail for each product.

(e) Cross-flow filtration systems can have relatively complex pipe work systems and the set up and cleaning of these must be studied in detail for impact upon production schedule.

(f) As scale increases, viscosity changes with concentration must be studied to ensure heat and mass transfer properties are not affected. As viscosity increases, product loss in pipe work will increase and flushing should be considered to maximize product recovery.

1.3.5 Extraction and precipitation

Extraction and precipitation are widely used throughout biotechnology within the primary recovery section of processes. The operations are heavily dependent upon heat transfer, mass transfer, and fluid dynamics and these parameters should be studied in detail as scale increases.

Factors to be considered during scale-up include:

(a) Scale-up and scale-down experimentation is essential in designing larger scale equipment. This must include an understanding of the rate determining step, be it mixing or reaction.

(b) The method of chemical addition can be critical if mass transfer is the rate determining step. Poor distribution can result in local precipitation and loss of product. In-line mixers, agitator shaft, and multiple addition ports may need to be considered.

(c) The step should be optimized with the whole process in mind. Recovery should be compromised if the resultant precipitate is easier to handle downstream and produces a higher overall yield.

1.3.6 Utilities

As scale increases, the contribution of utilities towards the total product cost increases significantly. It is essential that the appropriate qualities of utilities are utilized throughout the process, with quality increasing with product purity. A classic example would be water usage where quality may increase from towns water through purified water to water for injection, as the product passes through purification.

1.3.7 Raw materials

As the process scales up, the quantities of raw materials will increase. As one moves from laboratory reagents to commercial bulk materials, there will often be a change in material specification. This may have an impact upon process operations. It is important that a clear understanding of the impact of scale on material quality is built into the process from day one. It is common practice to use the same grade of raw material for clinical trial work as will be used for commercial manufacture, even if this necessitates bulk ordering of material.

As scale increases, the design implications of solids handling activities increase. As far as possible, process additions, such as precipitants and buffer salts, should be added in liquid form. Raw materials should if possible be brought in as bulk liquids or, if this is not possible, be made up in liquid form in a dedicated area separated from any clean room operations.

1.3.8 Process solutions

At small scale, process solutions will generally be made up on a batch-by-batch basis. As throughput and scale increase this will involve an increasing number of labour-intensive make up, storage, and distribution operations. To simplify these operations the use of concentrated buffer operations should be considered. Bulk lots of concentrated buffer can be formulated and then diluted on an as required basis for process use. The concept can be extended to encompass fully automated in-line dilution systems but it must be stressed that such activities need to be assessed on a case-by-case basis. Critical issues in such an assessment include the number of buffers required and the tightness of acceptance criteria on buffer specification. The use of in-line operations is particularly attractive for simple salt solutions and for processes where buffers can be standardized across unit operations.

In certain industrial processes, the use of fully automated make up and distribution has been considered. Although the paper benefits of such an operation are attractive, the potential complexity of the engineering solution and attendant reliability must always be kept in mind when making design decisions.

1.4 Case study 1

This study involves the scale-up of a 5 litre laboratory process to a pilot plant operation, with consideration of final manufacturing scale. The aim of this study is to provide an overview of the development of scale-up strategy.

Table 4 Feedstock processing per quarter for case study 1

Period	Toxicity studies	Phase I trial	Phase II trial	Phase III trial	Full production
Quarterly demand (m^3)	0.35	0.4	1	3	83

Table 4 provides initial estimates of the quarterly volume of feedstock to be processed, which, when combined with the schedule for trials, encompasses steps 1 and 2 as described in Section 1.1.

The next step was to assess the best practicable scale of operation (see step 3 Section 1.1). Early in the development activities it was agreed that only two scale-up steps should take place: once for toxicity and clinical trails and a second to satisfy final market demand. The company concerned did not have any in-house production facilities and needed to decide whether to out source production or build a dedicated facility. The product is high volume and utilizes novel production and extraction technology. It was decided that the construction of a multi-product pilot plant was the best way to deliver production capacity for this and other products in the pipeline. The pilot plant design focused upon process operations at 50–250 litres scale and a 100 litre feed stock batch was chosen for toxicity and clinical trials. To meet the increasing demand in clinical trials the intensity of production would be increased from one batch every two weeks up to three or four batches each week for phase III trials. The initial basis for commercial production was set at a 1000 litre batch once per day. The scenario was ideal for two reasons: the scales were sufficient to provide all necessary scale-up scale-down data on operations, and the increasing intensity of operations allowed the company to develop robust operating philosophies for commercial production. In addition it was recognized that in moving to one batch each day, production in the pilot plant could allow generation of market entry product which could delay the need for investment in the commercial facility.

The step-by-step approach to selecting scale of operation can be summarized as follows:

(a) Build flexible cGMP pilot plant facility.

(b) Scale-up process from 10 to 100 litre scale.

(c) Increase process intensity to meet late stage clinical trial demand.

(d) Build commercial production facility based upon 1000 litre batch size.

1.5 Case study 2 (5)

Delta Biotechnology Ltd. has developed a process for the production of therapeutic recombinant human albumin from yeast, Recombumin albumin. In order to compete with albumin products from human blood plasma fractionation, the production facilities will be significantly larger than traditionally required for recombinant protein products. The production demands of such a process have resulted in a process that ensures that high throughput of material is possible. The case study below describes some of the issues that arose when the process was scaled-up from the development pilot plant to a pilot production facility.

Post-fermentation, a centrifugation step is performed to separate the cells from the extracellular product. Downstream of cell separation the product is purified through a multi-stage chromatographic process. This process utilizes positive and negative chromatography steps, radial and axial flow chromatography columns, and ultrafiltration. After purification there is a final ultrafiltration step followed by product formulation and filling.

This case study will briefly examine some of the scale-up issues associated with cell separation, radial flow chromatography, and process raw materials.

1.5.1 Cell separation

Duty. Recovery of extracellular recombinant protein product from yeast culture broth. Yeast is removed as a waste stream while the product is retained in the centrate.

Production scale unit operation. Centrifugation using a continuous disc-stack centrifuge with nozzle discharge.

The problem. The final production scale process would utilize a different type of centrifuge to that available at laboratory or pilot scale. There were concerns that process development work may be done on centrate which had a different profile with respect to host organism derived impurities:

(a) Initial process development was done at laboratory scale using bucket centrifuges in batchwise operation to produce very efficient removal of process organism and debris and hence centrate, with a lower impurity profile than that from production scale centrifugation.

(b) Development work was then scaled-up. The pilot plant used a continuous disc-stack centrifuge with intermittent discharge. The nozzle discharge feature is only available on much larger machines. The high differential pressures experienced during discharge led to higher levels of cell breakage and debris than in the production scale centrifugation.

(c) Production plant used continuous disc-stack centrifuge with nozzle discharge. The mechanism of discharge is gentler than the pilot scale intermittent discharge machine giving lower levels of cell breakage and debris levels.

(d) Centrate polishing filtration is required at pilot and production scale to remove the remaining low levels of the production organism and to remove debris, which would otherwise block chromatography matrices used in downstream processing. This was not required in the laboratory scale process because of the higher centrifugation efficiency.

(e) Differences were observed in the chromatographic performance of material generated by the three different centrifugation methods because of the different impurity profiles. A study was undertaken to investigate why pilot and production facility recoveries were lower than those obtained in the laboratory.

(f) It is practically impossible to obtain the same quality of centrate at all scales because of fundamental equipment design issues associated with batch versus continuous operation and with different types of discharge mechanism.

(g) It is not possible to scale-down the performance of a large production machine into the laboratory. Subsequent laboratory process optimization work used centrate samples from the production plant as feedstock to ensure that the performance of the laboratory scale downstream processing was representative of that at large scale.

1.5.2 Radial flow chromatography (RFC)

Perceived advantages of RFC are:

(a) Larger cross-sectional area for flow therefore lower pressure drops than axial flow columns.

(b) Ability to pump pack resulting in increased reproducibility of packing and reduced matrix exposure to the environment.

(c) Reduction in floor space occupied, compared to similar axial flow columns.

Laboratory scale RFC columns have shorter bed depths than at production scale and are unsuitable for development work. Axial flow columns were thus used in the laboratory and RFC in the development pilot plant. Wedge columns are now available to mimic production scale RFC in the laboratory for process development. A wedge column is a 100 ml column with a horizontal flow path of similar length to production scale columns. The vertical walls of the column converge towards the end of the column, mimicking the changing surface area that exists in production scale columns. The following issues should be considered:

(a) Understanding of variable linear velocity within the column bed is important because of the changing surface area that exists in production scale columns. Wedge columns can now be used for scale-down experiments. The impact on process dynamics must be understood.

(b) Scale-up on constant volumetric flow rate was recommended by column manufacturer and found to be satisfactory.

(c) Pressure flow curves from Wedge columns have a similar profile to pressure flow curves from 100 litre production scale columns. This gave confidence that Wedge columns are useful tools for scale-down experimentation.

The process was scaled-up from 10 litre development columns to 100 litre production columns. Individual 100 litre columns have been operated for over 300 cycles without the need for repacking.

1.5.3 Process raw materials

During the development of the process it was critical that commercial grade raw materials were used to ensure consistency of product qualities at all scales, as well as consideration of the scale-up of raw materials handling.

(a) Investigate to ensure that quantities are available at the proposed operational scale. Analar and other laboratory grade materials are generally prohibitively expensive in large quantities, if available, although convenient during the small scale development.

(b) Small quantities of commercial raw materials are not always available and it may be necessary to purchase much larger quantities than required.

(c) Specifications to be in place at the earliest stage to ensure traceable development work for regulatory submissions across all scales; these specifications should include microbial and endotoxin specifications where appropriate to the process.

(d) Quality audit of potential suppliers is necessary to ensure compliance with cGMPs.

(e) Use of liquid constituent acid/bases versus solid forms. The handling of materials in liquid form is more convenient at large scale, e.g. sodium hydroxide solution and acetic acid are used to make sodium acetate.

(f) Materials specifications with respect to concentration range can be critical if trying to make up buffers directly with raw material solutions (e.g. sodium hydroxide solution). It may be necessary to dilute to specific concentration (in-house or by third party).

(g) The 'insolubles' content of some raw materials can cause filter blinding, e.g. tetraborate solutions. The extra filtration costs should be factored into the cost comparisons. There can be large batch-to-batch variabilities, and it may be necessary to purchase pre-tested batch-specific quantities in conjunction with suppliers.

References

1. US Food & Drug Administration, Department of Health & Human Services 21 CFR parts 600–799 (Biological Products for Human Use), US Federal Register, 1998.
2. US Federal Register November 30. (1995). Vol. 60, No. 230.
3. Keshavarz-Moore, E. (19XX). In *Downstream processing of natural products* (ed. M. Verral). John Wiley, ISBN 0 471 963267, 1996.
4. Wallworth, D. M. (19XX). In *Downstream processing of natural products* (ed. M. Verral). John Wiley, ISBN 0 471 963267, 1996.
5. Freestone, R., Davis, C., and Wigley, A. Personal communication, Centeon, 1997.

List of suppliers

Anderman and Co. Ltd., 145 London Road, Kingston-upon-Thames, Surrey KT2 6NH, UK.
Tel: 0181 541 0035   Fax: 0181 541 0623

Beckman Coulter (UK) Ltd., Oakley Court, Kingsmead Business Park, London Road, High Wycombe, Buckinghamshire HP11 1JU, UK.
Tel: 01494 441181
Fax: 01494 447558
URL: http://www.beckman.com
Beckman Coulter Inc., 4300 N Harbor Boulevard, PO Box 3100, Fullerton, CA 92834-3100, USA.
Tel: 001 714 871 4848
Fax: 001 714 773 8283
URL: http://www.beckman.com

Becton Dickinson and Co., 21 Between Towns Road, Cowley, Oxford OX4 3LY, UK.
Tel: 01865 748844   Fax: 01865 781627
URL: http://www.bd.com
Becton Dickinson and Co., 1 Becton Drive, Franklin Lakes, NJ 07417-1883, USA.
Tel: 001 201 847 6800
URL: http://www.bd.com

Bio 101 Inc., c/o Anachem Ltd., Anachem House, 20 Charles Street, Luton, Bedfordshire LU2 0EB, UK.
Tel: 01582 456666
Fax: 01582 391768
URL: http://www.anachem.co.uk

Bio 101 Inc., PO Box 2284, La Jolla, CA 92038-2284, USA.
Tel: 001 760 598 7299
Fax: 001 760 598 0116
URL: http://www.bio101.com

Bio-Rad Laboratories Ltd., Bio-Rad House, Maylands Avenue, Hemel Hempstead, Hertfordshire HP2 7TD, UK.
Tel: 0181 328 2000
Fax: 0181 328 2550
URL: http://www.bio-rad.com
Bio-Rad Laboratories Ltd., Division Headquarters, 1000 Alfred Noble Drive, Hercules, CA 94547, USA.
Tel: 001 510 724 7000
Fax: 001 510 741 5817
URL: http://www.bio-rad.com

Calbiochem, San Diego, CA, USA.

CP Instrument Co. Ltd., PO Box 22, Bishop Stortford, Hertfordshire CM23 3DX, UK.
Tel: 01279 757711
Fax: 01279 755785
URL: http://www.cpinstrument.co.uk

Dupont (UK) Ltd., Industrial Products Division, Wedgwood Way, Stevenage, Hertfordshire SG1 4QN, UK.
Tel: 01438 734000
Fax: 01438 734382
URL: http://www.dupont.com

Dupont Co. (Biotechnology Systems Division), PO Box 80024, Wilmington, DE 19880-002, USA.
Tel: 001 302 774 1000
Fax: 001 302 774 7321
URL: http://www.dupont.com

Eastman Chemical Co., 100 North Eastman Road, PO Box 511, Kingsport, TN 37662-5075, USA.
Tel: 001 423 229 2000
URL: http://www.eastman.com

Fisher Scientific UK Ltd., Bishop Meadow Road, Loughborough, Leicestershire LE11 5RG, UK.
Tel: 01509 231166
Fax: 01509 231893
URL: http://www.fisher.co.uk
Fisher Scientific, Fisher Research, 2761 Walnut Avenue, Tustin, CA 92780, USA.
Tel: 001 714 669 4600
Fax: 001 714 669 1613
URL: http://www.fishersci.com

Fluka, PO Box 2060, Milwaukee, WI 53201, USA.
Tel: 001 414 273 5013
Fax: 001 414 2734979
URL: http://www.sigma-aldrich.com
Fluka Chemical Co. Ltd., PO Box 260, CH-9471, Buchs, Switzerland.
Tel: 0041 81 745 2828
Fax: 0041 81 756 5449
URL: http://www.sigma-aldrich.com

Hybaid Ltd., Action Court, Ashford Road, Ashford, Middlesex TW15 1XB, UK.
Tel: 01784 425000
Fax: 01784 248085
URL: http://www.hybaid.com
Hybaid US, 8 East Forge Parkway, Franklin, MA 02038, USA.
Tel: 001 508 541 6918
Fax: 001 508 541 3041
URL: http://www.hybaid.com

HyClone Laboratories, 1725 South HyClone Road, Logan, UT 84321, USA.
Tel: 001 435 753 4584
Fax: 001 435 753 4589
URL: http://www.hyclone.com

Invitrogen Corp., 1600 Faraday Avenue, Carlsbad, CA 92008, USA.
Tel: 001 760 603 7200
Fax: 001 760 603 7201
URL: http://www.invitrogen.com
Invitrogen BV, PO Box 2312, 9704 CH Groningen, The Netherlands.
Tel: 00800 5345 5345
Fax: 00800 7890 7890
URL: http://www.invitrogen.com

Life Technologies Ltd., PO Box 35, Free Fountain Drive, Incsinnan Business Park, Paisley PA4 9RF, UK.
Tel: 0800 269210
Fax: 0800 838380
URL: http://www.lifetech.com
Life Technologies Inc., 9800 Medical Center Drive, Rockville, MD 20850, USA.
Tel: 001 301 610 8000
URL: http://www.lifetech.com

Mauri Pinnacle Yeasts, Hull, MSE, Crawley, Sussex, UK.

Merck Sharp & Dohme, Research Laboratories, Neuroscience Research Centre, Terlings Park, Harlow, Essex CM20 2QR, UK.
URL: http://www.msd-nrc.co.uk
MSD Sharp and Dohme GmbH, Lindenplatz 1, D-85540, Haar, Germany.
URL: http://www.msd-deutschland.com

Millipore (UK) Ltd., The Boulevard, Blackmoor Lane, Watford, Hertfordshire WD1 8YW, UK.
Tel: 01923 816375
Fax: 01923 818297
URL: http://www.millipore.com/local/UK.htm

Millipore Corp., 80 Ashby Road, Bedford, MA 01730, USA.
Tel: 001 800 645 5476
Fax: 001 800 645 5439
URL: http://www.millipore.com

New England Biolabs, 32 Tozer Road, Beverley, MA 01915-5510, USA.
Tel: 001 978 927 5054

Nikon Inc., 1300 Walt Whitman Road, Melville, NY 11747-3064, USA.
Tel: 001 516 547 4200
Fax: 001 516 547 0299
URL: http://www.nikonusa.com
Nikon Corp., Fuji Building, 2-3, 3-chome, Marunouchi, Chiyoda-ku, Tokyo 100, Japan.
Tel: 00813 3214 5311
Fax: 00813 3201 5856
URL: http://www.nikon.co.jp/main/index_e.htm

Nycomed Amersham plc, Amersham Place, Little Chalfont, Buckinghamshire HP7 9NA, UK.
Tel: 01494 544000   Fax: 01494 542266
URL: http://www.amersham.co.uk
Nycomed Amersham, 101 Carnegie Center, Princeton, NJ 08540, USA.
Tel: 001 609 514 6000
URL: http://www.amersham.co.uk

Perkin Elmer Ltd., Post Office Lane, Beaconsfield, Buckinghamshire HP9 1QA, UK.
Tel: 01494 676161
URL: http://www.perkin-elmer.com

Pharmacia Biotech (Biochrom) Ltd., Unit 22, Cambridge Science Park, Milton Road, Cambridge CB4 0FJ, UK.
Tel: 01223 423723   Fax: 01223 420164
URL: http://www.biochrom.co.uk

Pharmacia and Upjohn Ltd., Davy Avenue, Knowlhill, Milton Keynes, Buckinghamshire MK5 8PH, UK.
Tel: 01908 661101
Fax: 01908 690091
URL: http://www.eu.pnu.com

Pierce Chemicals, Rockford, IL, USA.

Promega UK Ltd., Delta House, Chilworth Research Centre, Southampton SO16 7NS, UK.
Tel: 0800 378994   Fax: 0800 181037
URL: http://www.promega.com
Promega Corp., 2800 Woods Hollow Road, Madison, WI 53711-5399, USA.
Tel: 001 608 274 4330
Fax: 001 608 277 2516
URL: http://www.promega.com

Qiagen UK Ltd., Boundary Court, Gatwick Road, Crawley, West Sussex RH10 2AX, UK.
Tel: 01293 422911   Fax: 01293 422922
URL: http://www.qiagen.com
Qiagen Inc., 28159 Avenue Stanford, Valencia, CA 91355, USA.
Tel: 001 800 426 8157
Fax: 001 800 718 2056
URL: http://www.qiagen.com

Roche Diagnostics Ltd., Bell Lane, Lewes, East Sussex BN7 1LG, UK.
Tel: 01273 484644
Fax: 01273 480266
URL: http://www.roche.com
Roche Diagnostics Corp., 9115 Hague Road, PO Box 50457, Indianapolis, IN 46256, USA.
Tel: 001 317 845 2358
Fax: 001 317 576 2126
URL: http://www.roche.com
Roche Diagnostics GmbH, Sandhoferstrasse 116, 68305 Mannheim, Germany.
Tel: 0049 621 759 4747
Fax: 0049 621 759 4002
URL: http://www.roche.com

Schleicher and Schuell Inc., Keene, NH 03431A, USA.
Tel: 001 603 357 2398

Shandon Scientific Ltd., 93-96 Chadwick Road, Astmoor, Runcorn, Cheshire WA7 1PR, UK.
Tel: 01928 566611
URL: http://www.shandon.com

Sigma-Aldrich Co. Ltd., The Old Brickyard, New Road, Gillingham, Dorset XP8 4XT, UK.
Tel: 01747 822211
Fax: 01747 823779
URL: http://www.sigma-aldrich.com
Sigma-Aldrich Co. Ltd., Fancy Road, Poole, Dorset BH12 4QH, UK.
Tel: 01202 722114
Fax: 01202 715460
URL: http://www.sigma-aldrich.com
Sigma Chemical Co., PO Box 14508, St Louis, MO 63178, USA.
Tel: 001 314 771 5765
Fax: 001 314 771 5757
URL: http://www.sigma-aldrich.com

Stratagene Inc., 11011 North Torrey Pines Road, La Jolla, CA 92037, USA.
Tel: 001 858 535 5400
URL: http://www.stratagene.com
Stratagene Europe, Gebouw California, Hogehilweg 15, 1101 CB Amsterdam Zuidoost, The Netherlands.
Tel: 00800 9100 9100
URL: http://www.stratagene.com

United States Biochemical, PO Box 22400, Cleveland, OH 44122, USA.
Tel: 001 216 464 9277

Westfalia Milton Keynes, UK.

Wheaton, NJ, USA.

Index